ABOUT ISLAND PRESS

Island Press is the only nonprofit organization in the United States whose principal purpose is the publication of books on environmental issues and natural resource management. We provide solutions-oriented information to professionals, public officials, business and community leaders, and concerned citizens who are shaping responses to environmental problems.

In 1994, Island Press celebrated its tenth anniversary as the leading provider of timely and practical books that take a multidisciplinary approach to critical environmental concerns. Our growing list of titles reflects our commitment to bringing the best of an expanding body of literature to the environmental community throughout North America and the world.

Support for Island Press is provided by The Geraldine R. Dodge Foundation, The Energy Foundation, The Ford Foundation, The George Gund Foundation, William and Flora Hewlett Foundation, The John D. and Catherine T. MacArthur Foundation, The Andrew W. Mellon Foundation, The Joyce Mertz-Gilmore Foundation, The New-Land Foundation, The Pew Charitable Trusts, The Rockefeller Brothers Fund, The Tides Foundation, Turner Foundation, Inc., The Rockefeller Philanthropic Collaborative, Inc., and individual donors.

SHAPING NATIONAL RESPONSES TO CLIMATE CHANGE

SHAPING NATIONAL RESPONSES TO CLIMATE CHANGE

A POST-RIO GUIDE

EDITED BY HENRY LEE

ISLAND PRESS

Washington, D.C. • Covelo, California

Library of Congress Cataloging in Publication Data

Shaping national responses to climate change: a post-Rio guide/
    edited by Henry Lee.
        p.   cm.
    Includes bibliographical references and index.
    ISBN 1-55963-343-3. — ISBN 1-55963-344-1 (pbk.)
    1. Environmental policy.   2. Climatic changes.   3. Greenhouse
gases—Environmental aspects.   4. Greenhouse gases—Political
aspects.   5. United Nations Framework Convention on Climate Change
(1992)   I. Lee, Henry.
GE170.S48   1995
363.73'87—dc20   94-39851
                                                      CIP

Printed on recycled, acid-free paper ⊚ ✸

Manufactured in the United States of America

10   9   8   7   6   5   4   3   2   1

■ CONTENTS

■ CONTRIBUTORS

ABRAM CHAYES is Felix Frankfurter Professor of Law Emeritus, Harvard Law School. Address: Harvard Law School, Cambridge, MA 02138.

ROBERT W. HAHN is Resident Scholar, American Enterprise Institute, 1150 17th Street, Washington, D.C. 20036.

DAVID HART is Assistant Professor of Public Policy, John F. Kennedy School of Government, Harvard University. Address: John F. Kennedy School of Government, 79 JFK Street, Cambridge, MA 02138.

DALE W. JORGENSON is Frederic Eaton Abbe Professor of Economics, Harvard University, and Director, Program on Technology and Economic Policy, John F. Kennedy School of Government, Harvard University. Address: John F. Kennedy School of Government, 79 JFK Street, Cambridge, MA 02138.

HENRY LEE is Director of the Environment and Natural Resources Program, Center for Science and International Affairs, John F. Kennedy School of Government, Harvard University. He is also a lecturer at the John F. Kennedy School of Government. Address: John F. Kennedy School of Government, 79 JFK Street, Cambridge, MA 02138.

RONALD B. MITCHELL is Assistant Professor of Political Science, University of Oregon. Address: Department of Political Science, University of Oregon, Eugene, OR 97403.

VICKI NORBERG-BOHM is Assistant Professor of Environmental Policy and Planning, Massachusetts Institute of Technology. Address: Department of Urban Planning, Massachusetts Institute of Technology, Cambridge, MA 02139.

EDWARD A. PARSON is Assistant Professor of Public Policy, John F. Kennedy School of Government, 79 JFK Street, Cambridge, MA 02138.

JAMES K. SEBENIUS is Professor of Business Administration, Harvard Business School. Address: Harvard Business School, Cambridge, MA 01238.

ROBERT N. STAVINS is Associate Professor of Public Policy, John F. Kennedy School of Government, Harvard University. Address: John F. Kennedy School of Government, 79 JFK Street, Cambridge, MA 02138.

BRUCE N. STRAM is Vice President, Corporate Strategy and Planning, Enron Corporation. Address: Enron Corporation, P.O. Box 1188, Houston, TX 77251-1188.

RAYMOND VERNON is Clarence Dillon Professor of International Affairs Emeritus, John F. Kennedy School of Government, Harvard University. Address: John F. Kennedy School of Government, 79 JFK Street, Cambridge, MA 02138.

PETER J. WILCOXEN is Assistant Professor of Economics at the University of Texas at Austin. Address: Department of Economics, University of Texas, Austin, TX 78712.

RICHARD J. ZECKHAUSER is Frank Plumpton Ramsey Professor of Political Economy, John F. Kennedy School of Government, Harvard University. He is also Director, Harvard Faculty Project on Regulation. Address: John F. Kennedy School of Government, 79 JFK Street, Cambridge, MA 02138.

■ PREFACE

In June 1992, negotiators from more than 150 countries met in Rio de Janeiro and signed what is now known as the Framework Convention on Climate Change. By the time this volume is published, over 100 countries will have ratified the convention and the first meeting of the Conference of the Parties will have convened in Berlin.

By signing the Framework Convention, nations commit to work together to address a common problem. The task of negotiating, designing, developing, and implementing specific policies and programs to respond to that problem now begins—an undertaking that will extend well into the next decade and will demand a level of creativity and political acumen rarely matched in the annals of international environmental discussions.

Negotiators will be forced to grapple with issues at the core of nations' economies and social aspirations. Disparities in income, culture, national resources, and perceptions of fairness will make it very difficult to find common ground. For example, per capita energy consumption in countries of the Organization for Economic Cooperation and Development (OECD) is fifteen times higher than that in lower income economies. While energy growth in the former stabilized at 1 percent in the 1980s, that of low-income economies increased by 5.3 percent.[1] It will be almost impossible for negotiators to convince low-income countries to reduce their efforts to expand their economies and provide a higher standard of living for their people, yet a tripling or quadrupling of fossil fuel consumption in these countries will make it very difficult to reduce the growth in greenhouse gas emissions and the risk of global climate change.

With the signing of the Framework Convention, hundreds of experts and public officials in nearly one hundred countries will be engaged in a process to assess a broad menu of strategies to reduce the amount of greenhouse gases released into the atmosphere.[2] In the United States, the Climate Change Action Plan, announced by President Clinton in October 1993, contained forty-seven initiatives involving industry, transportation, housing, foreign policy, forestry, energy, environmental protection, and agriculture. This plan will be reviewed at least every two years. All other countries that ratify the Framework Convention will prepare and submit similar plans. These parallel processes—an international process through the Conference of the Parties and a domestic process through the development of national action plans—guarantee that design, development, and implementation of mitigation responses will be on the public policy agenda for many years.

The purpose of this book is to inform these processes. It lays out the factors that policymakers should consider in designing their responses. It explores paradigms and trade-offs, not merely single answers. Because the signing of the Framework Convention on Climate Change in June 1992 marked the beginning of an international negotiation process that will last for decades, the lessons in this book should be as relevant ten years from now as they are today. The early chapters focus on issues related to the negotiation of international environmental agreements—designing negotiation strategies, linking the politics of international negotiations with domestic policy development, and developing tactics that might be most effective in designing enforceable international initiatives.

The remaining chapters assess the effectiveness and efficiency of specific policies: carbon taxes, tradeable permits, and technology transfer initiatives. Many variations exist of each, and in four chapters it is difficult to include all of the relevant options. Instead, the authors identify the key factors that should be weighed in the design, development, and implementation of any policy that falls within the rubric of these three policy initiatives.

Many of the mitigation measures under active discussion involve the development and dissemination of technologies that either consume less energy or replace fossil fuels. Alternative vehicles, more efficient appliances, solar photovoltaic systems, wind power, and fuel cells are on many lists of promising future technologies. The authors of this volume agree that technology development and dissemination will play a critical, if not paramount, role in any mitigation strategy. Robert W. Hahn, Robert Stavins, Dale Jorgenson, Peter J. Wilcoxen, and Bruce Stram all argue that if governments judiciously use market incentives, such as carbon taxes or tradeable permits, they can create a market climate that will stimulate innovation leading to the development and dissemination of hundreds of new technologies. Therefore, this volume targets market incentives rather than particular emerging technologies.

Finally, this book is not Harvard University's response to the Clinton administration. The Clinton action plan ensures a dynamic process rather than a single plan. It will inevitably go through many iterations over the next decade. Policies and initiatives deemed politically untenable at one time may become acceptable at another. Our hope is that these chapters will provide frameworks—not single answers—for their design, development, and implementation of these.

NOTES

1. *Far Eastern Economic Review*, 18 June 1992, 50.
2. For the Framework Convention on Climate Change to take effect, fifty countries must ratify it. At the conference in Rio de Janeiro in 1992, 161 countries signed the agreement.

■ ACKNOWLEDGMENTS

This volume represents the culmination of a four-year research project, the Harvard Global Environmental Policy Project, which involved thirty to forty scholars and experts. Many more people deserve to be acknowledged than I have space in which to do so. However, my coauthors and I would like to single out a few.

The Harvard Global Environmental Policy [Program] was originally designed and developed by Professor William Clark, director of Harvard University's Center for Science and International Affairs, and Florence Fisher, then assistant director of the Environment and Natural Resources Program. Without their help and guidance, this project would never have begun.

Many of the ideas contained in the chapters on negotiation emerged from discussions conducted under the auspices of the Harvard Negotiation Roundtable, chaired by Professor Howard Raiffa. Both he and the participants in the Roundtable played a critical role in creating the intellectual environment for new ideas. One of the most active participants was Professor Thomas Schelling. All of us who worked with Tom realize that our discussions were much richer as a result of his involvement. Others who participated in these discussions include Arthur Applebaum, William Moomaw, Richard Benedict, Richard Vietor, and David Victor.

Any project of this type would not take place without the support of the community in which it is conducted. We owe thanks to past and present deans of the John F. Kennedy School of Government—Graham Allison, Robert Putnam, and Albert Carnesale—and to the former director of the Center for Science and International Affairs, Ashton Carter. I especially want to express my gratitude to William Hogan and Harvey Brooks, both of whom supported this effort since its inception and served as sounding boards for our ideas.

Much of the research that went into these chapters was supported by a grant from the U.S. Environmental Protection Agency. We are especially grateful to Richard Morgenstern, Dennis Tirpak, and Alex Cristofaro for their confidence in our work and for their patience.

Finally, we would like to thank the people who helped edit, review, and prepare this volume. Miriam Avins and Teresa Pelton Johnson did an amazing job in transforming our stilted writing. Jo-Ann Mahoney, Jesselynn Pelletier, and Karen Rothschild put in many long hours to prepare the manuscript, organize it, and deal with a myriad of details. Carter Wall and Bill White served as my research assistants and showed amazing resilience in the face of hundreds of requests for obscure facts, citations, and data. Last but certainly not least, I am grateful to my wife, Mary, and my family for their support in this venture.

1 Introduction

■ **HENRY LEE**
John F. Kennedy School of Government
Harvard University

Modern communication, with its emphasis on ten-second sound bites, sensational headlines, and laser-sharp images, bombards today's society with a pictorial and verbal cacophony of sights and sounds—the Berlin Wall crumbling, Tiananmen Square exploding in revolution, the cries of children starving in African refugee camps. However, perhaps the most influential image of the past three decades is that of the planet Earth—a delicate blue orb, partially covered by wispy clouds. Yet as this image has been lodged in the public's psyche, so too has the awareness that the earth faces the possible threat of irreversible harm on an unprecedented scale. The ozone hole, the loss of biodiversity, and the population explosion are now all inextricably associated with that image. But the most difficult, complex, and, perhaps, uncertain of the problems threatening the planet is the possibility of significant and rapid climate change, commonly known as the greenhouse effect.

All of these environmental problems feed on one another. A growing population demands more energy, and more energy use increases greenhouse gas emissions, such as carbon dioxide and methane. More people need more arable land, which accelerates the pressure to clear-cut and burn forests, increasing the loss of biodiversity and the risk of climate change on a global scale. The fundamental issue is that humankind is altering, in ways that are not well understood, all of the systems and cycles that together make life on earth, as we know it, possible.[1] This book explores global climate change not only because it may prove to be one of the defining problems of the next century, but also because the processes and policies that will be effective in responding to this problem can be used to resolve other global environmental problems.

OUR FOCUS

Much of the recent debate on global warming has focused on whether the world is moving toward a rapid and calamitous descent into a very different and warmer climate regime or whether these fears are the product of incomplete science, media sensationalism, and a risk-averse populace. We leave these scientific issues to those who are more qualified to assess them and instead address the strategic policy issues that are related to designing, negotiating, implementing, and enforcing national responses to the problem of global climate change. The purpose of this book is to draw attention to alternative ways of thinking about how to design and gain political support, both domestically and internationally, for strategies and policies that could effectively reduce greenhouse gases.

Our point of departure is the report of the scientific working group of the Intergovernmental Panel on Climate Change (IPCC), which represented the best judgment of the world's scientific community at the time this book was written.[2] The thrust of the report is that over the next hundred years, the global mean temperature is likely to increase at rates faster than the world has experienced in the past and that this will affect both climate and climate-dependent ecosystems in ways that are uncertain but, potentially, very harmful. Some reputable scientists disagree with these findings, and future research may prove them to be correct.[3] Enormous scientific uncertainty surrounds this issue, and this uncertainty is not likely to evaporate any time soon.

Therefore, governments have three alternative courses of action, which can be pursued individually or in combination. The first option is for governments to wait until the level of scientific uncertainty is reduced, when there is more confidence in the cost-effectiveness of certain responses. Given the magnitude of the scientific uncertainties, there are strong arguments in favor of such a course. Such a strategy would emphasize research and would postpone programmatic investments, at least in the near term.

The second option is to emphasize adaptation, that is, to invest in programs and technologies that might permit countries to effectively adapt to the impacts of climate change—if and when they emerge. Such investments would lower the future cost of the possible impacts on forestry, agriculture, transportation, and economic infrastructure. There are strong economic efficiency and political arguments in favor of this second option. If credible responses to the threat of climate change take the form of societal investments in global insurance, adaptation investments may be a logical and reasonable down payment.

The third option is to invest in mitigation initiatives—initiatives to reduce greenhouse gas emissions or to increase carbon dioxide sinks. Most industrialized nations have decided to give priority to mitigation responses. This choice may change at a later date, and the extent of the initial mitigation investments may seem modest. For the foreseeable future, however, developed countries will discuss, and in many instances pursue, mitigation alternatives. For this reason, the authors of this volume focus on this option and seek to inform the choices governments will make as they pursue mitigation strategies on both the domestic and international fronts. Our focus is on selecting the goals, strategies, and programs to pursue and the tactics for designing, developing, and implementing them.

We start from the premise that the U.S. government, by signing the United Nations (UN) Framework Convention on Climate Change (FCCC), has concluded that it is in its national interest to reduce world greenhouse gas emissions. Our question is not whether the United States should act, but rather how an enforceable strategy can be fashioned that will be acceptable both domestically and internationally. We step back from the immediate debate and present a strategic framework to assess and compare specific alternatives. Many technological and programmatic options to reduce the threat of climate change have been suggested, but without political consensus or a strategy for building and institutionalizing such a consensus, no option will achieve its potential. Whether a nation believes that solar energy, energy conservation, or aggressive tree planting is the right answer, its first step must be the design of an overall strategy to realize its international and domestic objectives. Without such a coherent strategy, mitigation efforts are the strategic equivalent of shooting stars—spectacularly bright, but short-lived, political phenomena in the global political firmament.

This book brings together a distinguished group of scholars to explore what steps and factors should be considered in the development of response strategies and what might be the consequences of nations' adopting those strategies. Many of the chapters are written from the perspective of the United States. To some, this focus may seem parochial. U.S. policy prescriptions are not necessarily more insightful or visionary than those of other countries; however, as the only remaining world superpower, the United States plays a unique role in the global community of nations. We do not suggest that the United States embrace more aggressive or stringent reduction goals, but we recognize that whatever positions and policies the United States supports will have a substantial influence on the positions and policies taken by other nations.

The first half of this book explores the factors that shape the prospects of negotiators obtaining a workable international agreement. James K. Sebenius applies the experiences from previous international environmental negotiations to draw important lessons for the U.S. negotiators who will be given the task of developing specific initiatives and protocols to implement the framework convention signed at the 1992 UN Conference on Environment and Development (UNCED). Edward A. Parson and Richard J. Zeckhauser assess the critical problem of trying to craft quantitative measures of national performance—emissions targets or equivalent measures—which will remain a central problem in future negotiations. Ronald B. Mitchell and Abram Chayes focus on compliance and enforcement issues. No matter what targets are selected or what programs are endorsed, no treaty, protocol, or agreement will succeed unless the parties comply with the provisions. Finally, Raymond Vernon turns his attention to the importance of the linkages between the process by which domestic policy responses are developed and those through which international negotiating strategies are formulated. Such linkages are created (or not) by each country's history, value system, and institutions, and these elements can result in very different behavior patterns.

The second half of the book explores broad policies and programs that the United States could adopt domestically in response to a negotiated agreement to reduce greenhouse gas emissions. Robert W. Hahn and Robert N. Stavins explore the use of market-based incentives, specifically, tradeable permits, and suggest criteria for deciding when to adopt such programs and how to design them. Bruce N. Stram assesses the political and strategic merits of a carbon tax and provides specific recommendations as to how such a tax might be structured and sold to a skeptical public. Next, Dale W. Jorgenson and Peter J. Wilcoxen tackle the critical question of how much it will cost—in terms of both money and the impact on the economy—if the United States decides to adopt specific targets for carbon dioxide reduction. Finally, Vicki Norberg-Bohm and David Hart look at the lessons to be learned from past U.S. efforts at transferring new technologies to developing countries, highlighting the importance of thinking not only about what technologies to transfer, but also about how to deploy them most effectively.

This introductory chapter summarizes the fundamental lessons contained in the eight chapters that follow. I begin by briefly describing several of the scientific, political, and economic dimensions that shape how policymakers must approach the problem. I then synthesize the principal recommendations on how the United States might be able to forge a workable international agreement and explore specific policy paths that the United States should consider.

THE DIMENSIONS OF THE PROBLEM

Human activities, especially over the past hundred years, have added substantially to the concentration of greenhouse gases—such as carbon dioxide, methane, nitrous oxide, ozone, and chlorofluorocarbons (CFCs)— to the earth's atmosphere.[4] Since the industrial revolution, deforestation and the combustion of fossil fuels have led to a 26-percent increase in carbon dioxide in the atmosphere. Methane levels have doubled over this period.[5] If other environmental factors remain stable, and if the concentrations of greenhouse gases increase, the atmosphere's capacity to capture and hold heat will increase. This phenomenon should result in an increase in the earth's average temperature, which, in turn, could alter the level of precipitation, the rate of sea rise, and the number and severity of major storms. (See the appendix at the end of this volume for additional information and a summary of the principal findings of the scientific working group of the IPCC.)

Although this book does not examine the scientific debate, it is important to discuss a few aspects of the global climate change problem that will affect the magnitude and scope of governmental response. First, if the IPCC's scientific report is correct, it is unlikely that the world will be able to avoid some increase in average world temperature and perturbations in the form of changes in rainfall, storms, and other weather phenomena. Even if greenhouse gas emissions were radically reduced, substantial momentum has already been built into the system. Greenhouse gases have long atmospheric lifetimes. Carbon dioxide remains in the atmosphere for 50–200 years.[6] Nitrous oxide lasts for 150 years, and CFCs, 65–130 years. Only methane among the major man-made gases has a relatively short atmospheric lifetime, but even it lasts 10 years. The U.S. Environmental Protection Agency (EPA) calculated in a 1990 study that to stabilize atmospheric concentrations, a 75-percent reduction in greenhouse gas emissions would be required. In fact, the EPA estimated that carbon dioxide emissions would have to be reduced by 50 percent; nitrous oxides, by 80–85 percent; and methane, by 10–20 percent; CFCs would have to be phased out altogether.[7]

There is little likelihood that such emissions reductions will be realized. Yale University economist William Nordhaus noted in an article published in *Science* that a policy to stabilize greenhouse gas concentrations (that is, one that would hold the actual concentrations of gases in the atmosphere constant) would require investments of $30 trillion in discounted income (over the period 1985–2105) as compared with a no-controls policy.[8] The climate change treaty agreed to at the 1992 Earth Summit in Rio de Janeiro calls for nations to try to stabilize greenhouse gas emissions at 1990 levels. President Clinton committed the United States to this goal in April of 1993. The target may be laudable, but it will only minimally reduce the temperature increases projected by the IPCC, since concentrations, which are the governing factor, will continue to increase. Nordhaus estimated that the difference between a no-controls strategy and an emission-stabilization scenario may be less than three-tenths of one degree, albeit the difference increases over time and would be greater than 1 degree Celsius by the year 2105.[9]

The inference from these studies is that the levels of emissions reductions being discussed—even those suggested by the environmental community—will only slightly retard the increase in average world temperature, as well as the rate of those increases. Therefore, why commit billions of dollars to retard the growth in greenhouse gas emissions when the end result may be minimal?

There are two responses to this question. First, given all of the uncertainties, the experts may be underestimating the benefits from moderate reductions in greenhouse gases. Second, there is now evidence questioning the belief that changes in the world's climate will occur at rates sufficiently even and steady for the world to adapt with only moderate economic dislocations. Several studies of ice cores extracted from the Greenland ice sheet and Peru's Quelccaya ice cap, which date back 100,000–250,000 years, suggest that, historically, climate has shifted very rapidly.[10] That is, the earth's climate system resists change until pushed over a threshold, then it leaps into a new climate system. Dramatic change may come very suddenly. If so, adapting to change will prove to be much more difficult and the cost in human misery much higher. Further warming, in and of itself, is not the primary danger; rather it is perturbations in the world's weather system that might dramatically disturb the amount and location of rainfall and the intensity of storms.

Two scenarios capture the dilemma facing policymakers. In the first, the United States and other countries invest billions of dollars with little, if any, benefit. In the second, too little is invested, and the world's climates jump abruptly into a new equilibrium, with disastrous consequences. The high level of scientific uncertainty surrounding this issue makes it impossible for policymakers to avoid this dilemma. Professor

Harvey Brooks captured the essence of this policy choice when he wrote the following.

> Scientific uncertainty, both in the measurement of the current state and in the inferring of future trends from current data, is a major impediment to operationalizing the concept of sustainable development in both managed and natural ecosystems. In some cases, the insurance premium that may have to be paid in order to fully hedge against current technical uncertainties may be so high as to be economically unrealistic. Whether or not this premium is worth paying depends on the projected cost of reversing the future consequences of current decisions, if the premium is not paid.[11]

To think about the problem, consider a simple graph with time on one axis and temperature, as a surrogate for climate change, on the other. Assume that there is a correlation between the rate and magnitude of climate change and the damage to the planet's environment. There are certain trajectories that governments would find unacceptably costly, either because average global temperature rises too rapidly or because the absolute magnitude of the change is too great. The challenge for policy-makers is twofold: first, to ensure that the wrong trajectories do not occur and, second, to plan so that the correct trajectories are not achieved in an unacceptably costly and inefficient manner. These two challenges are of equal concern.

DESIGNING A U.S. STRATEGY

In the chapters that follow, the authors discuss many important issues. Some, such as the differences in interest and values between the developed and the developing worlds, constrain what can be done; others, such as the need to create new international institutions and workable enforcement and monitoring mechanisms, will be essential ingredients in any effective climate change agreements or protocols. There are, however, several fundamental elements of the global climate change problem that will have to be factored into the design of any mitigation response.

1. The problem could be very large.
2. Because of scientific and economic uncertainties, the perceptions of the problem will be in constant flux.

3. Efforts to effectively reduce greenhouse gas emissions will carry large up-front costs.

4. The development and dissemination of new and efficient technologies will be especially important.

5. Energy prices, especially those for carbon-intensive fuels, will have to increase.

First, the design of U.S. domestic policy options to mitigate global warming will be a much more complex task than was the design of responses to past environmental problems. The cost of even stabilizing emissions is several orders of magnitude larger than is the cost of phasing out chemicals that destroy the ozone layer or of reducing the threat of acid rain. Furthermore, climate change is a global commons problem—no nation can solve it alone, nor can any nation insulate itself from the consequences of another's actions. For example, recent commitments by the Organization for Economic Cooperation and Development's (OECD) member countries to reduce emissions by 2000 will almost certainly be offset by the growth in emissions in large, rapidly growing developing countries.[12]

Vernon makes the point that domestic policy cannot get too far in front of, or lag too far behind, positions taken internationally without undercutting the credibility and bargaining position of U.S. negotiators. This linkage between international and domestic policies is often ignored or not fully appreciated. International policy is formulated by the executive branch, with only moderate involvement by special interests, whereas domestic policy is a product of a pluralistic, interactive process involving the executive branch, the Congress, state and local governments, and a spectrum of interest groups. Thus, ideological and specific economic concerns, such as the impact on regions or on certain industries, tend to surface with more emotion during the formation of domestic policy. If the United States is to craft an effective global strategy for greenhouse gas mitigation, the processes by which it formulates domestic and international policies must be closely linked.

Second, since the scientific understanding of global climate change will be constantly evolving, international and domestic policy responses must be flexible. Over the next two decades, countries will go through five or six cycles of debating, developing, and implementing responses to the problem of global warming. Each cycle will be triggered by new scientific information. Policymakers should be careful to recognize that new information will not always point in the same direction and that each policy development cycle will build on the one which preceded it. In some instances, new information may contradict or repudiate past efforts.

There is a dangerous tendency to assume that the evolution of science and policy witnessed in the case of CFCs will be replicated in the case of global warming. In that case, the worst-case scenarios forecasted in the late 1970s and early 1980s came to pass, and there was a linear progression of evidence indicating that the problem was even more serious than anticipated. While in the case of climate change a scientific consensus, supported by a continuous flow of hard evidence, may emerge, there is a much stronger likelihood that the science will take a nonlinear path, characterized more by zigs and zags than by a straight line. Therefore, policy responses must be flexible. Because new information may suggest that more or less stringent reductions in greenhouse gas emissions are warranted, response mechanisms need to be sufficiently flexible so that they can be changed relatively quickly, without destroying basic institutional frameworks or processes. Flexibility is important to ensure the continuity of support and participation.

Third, the scale of the problem makes cost-effectiveness crucial. If the United States is to stabilize or reduce carbon dioxide or other greenhouse gas emissions, it cannot afford to duplicate some of the inefficient and costly programs that have characterized past U.S. responses to other environmental and energy problems. (Stram chronicles the results of several of these efforts.) Jorgenson and Wilcoxen, who assess the costs of stabilizing carbon dioxide emissions at 1990 levels and the costs of reducing emissions by 20 percent below those levels, find that a stabilization policy would reduce the growth in U.S. gross national product (GNP) in the year 2020 by little more than one-half of one percent, while a 20-percent reduction would result in a GNP level three times lower.[13] These costs may be politically feasible, but they are thirty to fifty times greater than the costs of phasing out CFCs. Furthermore, these estimates are based on the assumption that the United States will adopt a carbon tax rather than a reduction program, which would consist of less efficient and more expensive command-and-control regulation accompanied by targeted subsidies—the type of programs traditionally favored by the U.S. Congress.

Fourth, policy responses will not be effective if they depend entirely on countries foregoing economic growth or on consumers voluntarily reducing their quality of life. Moral persuasion has a place, and people will take actions to help the environment. However, the magnitude of the reduction required will make any significant emissions reductions impossible to achieve without concomitant advances in the technical capacity to abate greenhouse gas emissions, to provide affordable energy alternatives that will emit less carbon dioxide, or to create carbon dioxide sinks in the form of large-scale reforestation projects. Technology development will be important to obtain the support of developing countries, especially those with indigenous coal resources, rapidly expanding economies, and large populations. Barring advances in alternative energy technology, these countries will use their coal resources at ever-increasing rates.

An effective policy must provide strong and stable incentives to encourage technological development. Governments should not preselect a limited number of alternative technologies and subsidize them; future developments may prove those selections to be unacceptable—either economically or culturally. Rather, governments should establish incentives that will stimulate the creativity of thousands of entrepreneurs and investors to develop more environmentally friendly technologies and link them to the needs of the marketplace.

Fifth, the U.S. government cannot continue to vigorously defend the right of the American consumer to enjoy low energy prices and expect to make any real progress in reducing carbon dioxide emissions. It may be perfectly appropriate for the U.S. government to take the position that low gasoline prices and cheap electricity are more important to the future well-being of the United States than are the benefits from reduced carbon dioxide emissions. However, to suggest that the U.S. government must take a leadership position on combating global warming, while it simultaneously resists domestic increases in energy taxes, is a position that lacks credibility. Advocating a forty-miles-per-gallon efficiency standard for new cars while fighting to keep gasoline prices below those in 1972 (when inflation is factored out) sends conflicting signals about the U.S. government's commitment.

OBTAINING A WORKABLE INTERNATIONAL AGREEMENT

Summarizing the principal lessons of the chapters on forging a workable international agreement is a daunting task, given their richness. Yet there are several areas in which the recommendations and insights of the authors are at variance with popular opinion. Unfortunately, some of these opinions are based on misconceptions. Four of these are especially important.

Popular opinion: An international consensus exists for a strong global climate change treaty.

Reality: No significant worldwide consensus for action exists. It must be forged.

Many environmental advocates believe that the opposition of key White House aides in the Bush administration was responsible for weakening the framework convention and for undercutting the negotiations leading to the Rio summit. The truth is that, currently, support for significant reductions in greenhouse gases among the nations of the world ranges from slight to nonexistent. The global climate change issue involves both scientific uncertainty and large economic costs. Mitigation responses must be paid for in the near term by identifiable entities, while their benefits are uncertain and dispersed and will be realized far in the future. For these reasons, a sophisticated and concerted political effort is needed to build a constituency both within the United States and abroad.

Over the last two decades, presidents of the United States have suggested small taxes or fees as part of a package either to reduce U.S. vulnerability to oil disruptions or to reduce budget deficits.[14] Each time, there has been a strong public consensus that a serious problem existed, and yet in each instance a firestorm of protest erupted. A country determined to maintain gasoline prices at one-third the average price in Europe will not easily accept the higher energy prices that a meaningful greenhouse gas reduction plan will require.

As Vernon notes, U.S. negotiators have a long history of taking an activist line at the international negotiating table (or during the daily press conference), only to later find their positions undercut by a Congress responding to interest-group pressures they had ignored. A negotiating strategy that puts a premium on opportunistic innovation is not always the most effective tactic in a pluralistic democracy.

Despite the strong rhetoric voiced by European politicians before the Earth Summit, the European Community (EC), now the European Union (EU), was unable to negotiate an allocation for carbon dioxide emissions reductions among its members. Even more striking was the saga of the proposed European carbon tax that evolved into an energy tax, which became an energy tax that exempted energy-intensive industries and major exporters, which then became a tax conditional on comparable action by the EU's major trading partners and was finally tabled in December 1994.[15] As Sebenius points out, this experience is eloquent testimony to the power of blocking coalitions and to the present lack of an intense desire on the part of interest groups in some countries to see meaningful policies enacted.

If the consensus for action within industrialized nations is easily shattered, the task of persuading the developing world is even more daunting. It is not difficult to convince countries to sign a framework convention that does not contain specific commitments or timetables; it is far more difficult to convince Saudi Arabia to support meaningful reductions in the use of fossil fuels, or Brazil to reverse the deforestation of its western and northern states, or China to dramatically restructure its industrial development goals.

If U.S. policymakers wish to develop a strong international program, they must first develop a sophisticated long-term strategy to build a much stronger consensus for action. Substantially more patience and subtlety will be required than has been demonstrated in the past. Public concern will rise and fall over the next few decades as new scientific information becomes available. An effective effort to build support must be able to weather both the ups and downs of public opinion and will take decades, not months or years, to shape.

However, we do not advocate waiting until all the nations of the world complete massive domestic education campaigns. We do believe that aggressive constituency-building efforts must be undertaken in tandem with the development of negotiation and mitigation responses. The chapters by Mitchell and Chayes and by Sebenius argue that the negotiating process should not be regarded simply as a vehicle to reach formal binding agreements, but also should be viewed as a way to generate publicity, raise awareness, mobilize opinion, forge coalitions across national boundaries, and provide support for broad-based research. It would be unproductive for the negotiations process and the mitigation responses emanating from that process to get too far ahead of public opinion. Such a scenario is a prescription for backlash and failure.

It is difficult to ask public officials who have four-year job horizons to embrace a strategy that will take twenty years to complete. The U.S. government has been able to take a long-term perspective on issues such as arms control, but in those instances there was a consensus about the nature and magnitude of the threat, fewer special interests, and less scrutiny by the popular media. Global climate change makes a good story for the popular press and puts pressure on public officials to respond to these stories, even if the response is not conducive to the shaping of a long-term solution. Unfortunately, only a long-term strategy is likely to convince the major greenhouse gas emitters of the world to measurably reduce their use of fossil fuels.

This book's recommendations on how the United States might build a global constituency focus on the three primary groups that will shape the debate: governments, the expert community, and nongovernmental organizations (NGOs).

GOVERNMENTS SHOULD SUPPORT AND EXPAND THE COUNTRY-REPORTING REQUIREMENTS OUTLINED IN THE FRAMEWORK CONVENTION

The framework convention signed in Rio de Janeiro requires each nation to file a report on the actions they are taking, or are proposing to take, to reduce greenhouse gas emissions.[16] As both the Sebenius and the Mitchell and Chayes chapters point out, this provision, if beefed up and adequately funded, can provide an important vehicle for involving and mobilizing public and private officials, public interest groups, and the expert community in each country. The process that each country establishes to develop these reports will require government ministers and the external interest groups linked to those ministries to become involved. In many cases, this involvement will trigger debate and stimulate new ideas and, sometimes, new programs. In almost all cases, involvement will dramatically expand the basic understanding of the issues among the participants and of the domestic political feasibility of possible responses.

The U.S. EPA, along with the environmental agencies from many other countries, is involved in an international process to inventory greenhouse gas emissions. This process represents a major step toward establishing a verifiable accounting system from which improvements can be measured. The United States should join with its allies to ensure that financial support for this effort is made available, that the numbers generated are accurate, and that the program is expanded to include as many nations as practical. To the extent possible, data collection procedures should be standardized to achieve comparability across countries.[17]

It is important to establish at the outset that compliance and enforcement issues must be included in the development of each report. At a minimum, these reports will provide an opportunity for interest groups outside of government to suggest alternative enforcement vehicles and institutions that might be adopted in the development of future action plans.[18] If a country has no regulatory program or enforcement capacity, the requirement may catalyze an internal debate about establishing such a capability.

PROMOTE TRANSNATIONAL COALITIONS

In the years leading up to the Earth Summit, hundreds of experts were participating in discussions, preparing position papers, doing research, and lobbying. Their input was reflected in almost every document that came out of the conference. For example, the early work of the Scientific Committee on Problems of the Environment, an arm of the International

Council of Scientific Unions, was instrumental in laying the groundwork for the IPCC's working groups, which, in turn, provided the UN's Intergovernmental Negotiating Committee (INC) with the substantive foundation with which to forge the FCCC.[19] The INC, in turn, has been instrumental in setting the agenda for the first meeting of the Conference of the Parties scheduled to take place in Berlin in March 1995.

There are numerous cases in which experts in different countries with similar interests formed de facto transnational coalitions, or what some refer to as "epistemic communities."[20] Often these coalitions have enormous influence on the policy positions that are eventually taken by international organizations and governments. Sebenius argues that nations should foster the continued development of these epistemic communities not only to stimulate new avenues of research, but also to help create greater opportunities for consensus building and coordinated action. If a team of scientists from the United States, Great Britain, and Sweden announces the discovery of evidence that the planet is warming, this information will not necessarily be credible in China, India, Brazil, or Saudi Arabia. If economists from Saudi Arabia, Nigeria, and Russia announce that their research shows that it is far cheaper to adapt to global warming than to cut back on the use of fossil fuels, Germany, the United States, or Japan will not necessarily rush to embrace this finding without undertaking a similar analysis. It is in the long-term interest of all parties to encourage the development of transnational groups of experts that are formed on the basis of professional interest, geography, or shared concerns.

Nurturing effective epistemic groupings sounds easy, but is not. Patience, significant financial support, and new institutions will be required. Five or ten years from now, such an investment could pay huge dividends, especially if new information points to the need to reduce emissions quickly. Without the establishment and sustenance of transnational communities, it could take several years to reach international consensus on the credibility of new information, not to mention agreement on whether and how it should be acted upon. Transnational coalitions will also be valuable when members seek to garner support from key interests in their home countries and can serve as an informal avenue linking influential experts from different countries with key government officials, the media, and the public. However, as Vernon points out, epistemic communities are more proficient at selling policies and programs and getting them adopted than at ensuring that these policies are implemented and enforced.

NGOs SHOULD BE SUPPORTED IN THEIR INVOLVEMENT WITH THE POLICY DEVELOPMENT PROCESS

NGOs have become a potent force in coalescing public support for international environmental agreements.[21] Over 1,400 NGOs participated in the Earth Summit. With modern global communication technologies, groups can quickly develop informal links to rally their constituents and influence domestic and international opinion. In most instances, these groups have considerably more mobility than do governments to take positions. They are able to easily form international coalitions of like-minded organizations, or, as Vernon calls them, "boundary-straddling communities." Mitchell and Chayes point out a number of ways that NGOs can use the media to increase public awareness and bring pressure to bear on governments, as well as on companies, that blatantly violate international accords. NGOs are better at pressuring government and industry to effectively implement and enforce agreements than are coalitions of experts.

There are, however, problems with environmental NGOs. Some are much more thoughtful and accurate in their use of scientific information than are others. Unfortunately, the press usually does not discriminate between these two groups and thus can pass on information and impressions to the public that may be inaccurate. Furthermore, environmental NGOs often focus on a single issue, which can distort and conceal difficult societal choices. For some people, especially in the developing world, the NGO communities' exclusive emphasis on environmental goals reflects the values of industrialized nations rather than the values of nations that are only now becoming industrialized. Finally, there are questions of who these groups represent. Donors are not necessarily synonymous with members, and in some organizations members have a minimal impact on shaping the policies and priorities of the organization.

Despite these problems, NGOs can make important contributions because they are independent from government. In a world with no binding requirements for greenhouse gas reductions, one of the few sources of pressure to achieve progress will be the ability of NGOs to rally public concern. Governments should not obstruct that capability, but rather should design programs and policies to take advantage of it, while recognizing and guarding against the possible deficiencies.

Popular opinion: The goal of U.S. policy should be to negotiate as soon as possible protocols setting rigid timetables and targets for the reduction of specific greenhouse gases.

Reality: Indirect and flexible strategies will be more successful than highly visible and aggressive responses.

One of the fundamental messages in every chapter of this book is that success should not be measured by paper mandates, rhetorical positions, or the number of protocols signed: success should be measured by results. Some people may be euphoric at the signing of a rigid and stringent carbon dioxide protocol, but they probably should trade this transcending moment for an alternative that, while less visible, achieves (rather than promises) greater reductions in greenhouse gas emissions.

In most instances, indirect strategies will be more successful than head-on attacks that require rigid carbon dioxide reductions at the outset. Over time, an indirect approach is more likely to prepare the way for successful direct mandates. As long as the cost of control is perceived as high and as long as those who must bear the cost are easily identifiable, blocking coalitions will rise up to fight direct measures. In contrast to popular perceptions, the more rigid the measure, the less likelihood there is of actually achieving the promised level of greenhouse gas reductions. Is it realistic to believe that China and India will agree to reduce carbon dioxide emissions to levels that would measurably impair their economic development plans? Sebenius warns that blocking coalitions—based either on economic self-interest or ideology—cannot be ignored. He further argues that there is a real danger of stimulating and exacerbating the North-South conflict, which has simmered just below the surface in previous international environmental negotiations. Industrialized nations should take care not to have future negotiations erupt into an ideological confrontation with the developing world. Rigid mandates could trigger such a reaction. The subsequent backlash would create inertia and might undermine the opportunity for real progress.

An effective strategy is one that creates incentives, rather than requirements, for domestic political coalitions to act. An effective strategy will be one that does not trigger the aggressive opposition of powerful blocking coalitions or threaten national priorities. Until there is a stronger popular consensus about the urgency of the need to respond to the problem of climate change, carrots will work better than sticks.

Each of the chapters in this book offers specific recommendations to move the negotiations forward. Three are especially important.

THE SUBJECT AND NATURE OF THE FIRST FEW PROTOCOLS SHOULD BE SELECTED ON STRATEGIC RATHER THAN ON IDEOLOGICAL OR THEORETICAL GROUNDS

The early protocols should target issues that will likely produce successful negotiations. For example, negotiators should not start with a protocol to reduce nitrous oxides from rice farming and expect China and Japan to sign, or try to obtain a worldwide consensus for simultaneous reductions of all greenhouse gases. Sebenius advises that early protocols should contribute joint gains to a broad-based group of adherent countries, while reducing the likelihood of stimulating opposing interests. In making such a selection, negotiators should carefully assess how the powerful political forces in each country will perceive a protocol's costs and benefits. If a strong protocol is signed but not ratified, implemented, or enforced, it will have diverted resources from protocols that may have had a much better chance of being accepted, thereby undermining the credibility of the process itself. It is essential to begin with successes, not failures.

Mitchell and Chayes urge that countries consider which economic actors and sectors are most likely to alter their behavior in response to national efforts to legislate, implement, and enforce treaty commitments. They present a list of determinants to identify such economic actors, which include large resources, high visibility, degree to which they currently face regulation, and ability to pass on incremental costs. Guided by this list, negotiators might focus more on protocols that would demand actions of large, visible multinational corporations, or certain state-owned enterprises, and less on protocols that would require the compliance of farmers or small, decentralized enterprises, which have less financial capabilities, are less visible, and face highly competitive markets.

PROGRESSIVE RATHER THAN RIGID TARGETS SHOULD BE SET

Both the Sebenius and the Parson and Zeckhauser chapters emphasize the importance of incremental approaches. The latter argues for the establishment of progressive obligations that are definable in terms of annual incremental improvements in trend lines for national emissions and carbon intensity, or even in terms of specific efforts, such as expenditures in emissions control or progressively increased tax levels for energy. Such an approach would respect nations' concerns about a precedent being set that might prove untenable and would also provide countries with maximum flexibility to review and adjust the annual incremental targets as they learn more about the costs and effectiveness of various policies and programs.

Sebenius suggests the use of ratchet mechanisms, whereby negotiators establish modest targets for greenhouse gas reductions and use these to institutionalize monitoring, enforcement, and implementation at both the international and domestic levels. Once the foundation is built, negotiators can simply ratchet up the targets if new information warrants more stringent reductions. Such a mechanism was a vital component of the Montreal Protocol for reducing CFC emissions.[22] A process characterized by many small steps rather than a few large ones has a much better chance of success.

THE NEGOTIATIONS PROCESS SHOULD BE INSTITUTIONALIZED

Negotiations to implement the FCCC and develop specific protocols will not be a two- or three-year effort. In fact, barring a definitive scientific finding that global warming will *not* occur, negotiators will be meeting on this topic well into the next century. Negotiations will be more akin to the ongoing discussions to amend the General Agreement on Tariffs and Trade (GATT), which have spanned decades, than to other international environmental negotiations, which often are completed in two to three years.

This book's recommendations assume that there will be continuous negotiations. In almost every case, the recommendations are designed to maximize the flexibility of each country to choose the means appropriate to its circumstances and to adjust to changes in scientific, political, technical, and economic factors. Furthermore, reporting requirements, such as those agreed upon at the Earth Summit, are a critical component of our recommendations, as they allow all parties to monitor the progress each country is making toward meeting its goals. An ongoing international mechanism to receive reports and negotiate adjustments is essential. How this mechanism should be structured and staffed is a question that should be answered sooner rather than later.

In recent years, the trend has been to maintain small central staffs, with most of the work being done by officials representing the national governments. The fear is that a larger staff would develop its own values and priorities and would attempt to impose them on the policy process. Governments are willing to cede some responsibilities to organizations that focus on problems deemed to be technical or scientific, such as the World Meteorological Organization. Governments are not so inclined if the organization's focus has a strong political component. Global climate change negotiation certainly falls within the latter category, and thus irrespective of administrative efficiency arguments, participating countries will likely insist on retaining responsibility for receiving and reviewing national action plans and for making most of the key decisions regarding the implementation of the framework convention.[23]

Popular opinion: To ensure fairness, an effective treaty should impose similar obligations on all parties.

Reality: A treaty that demands the same obligations from all nations will be impossible to negotiate or will be unsustainable once negotiated.

Parson and Zeckhauser point out that most international environmental treaties have asked all participants to assume identical burdens or responsibilities. Any differences have been highly constrained. For example, the Montreal Protocol imposed two targets—one for industrialized countries and a second for developing nations. Countries, however, differ dramatically in terms of wealth, population, geography, cost of mitigation, and how they value environmental quality over other social goals. Furthermore, the costs of environmental damage and the costs of reducing a nation's contributions to it do not fall equally on all people.

Despite these differences among nations, there are many reasons why negotiators of international environmental treaties have opted for what Parson and Zeckhauser term "symmetrical agreements." These simple agreements limit the informational burden on the negotiators, restrain value-claiming behavior (which is already a major problem in the global climate discussions), reduce the opportunities for negotiators to demand special consideration for their countries, and are usually easier to ratify. However, the costs of abating greenhouse gases differ so radically among countries that symmetrical solutions are, in the words of Parson and Zeckhauser, likely to be "impossible to negotiate or unsustainable once negotiated." The diversity of commitments among the OECD countries illustrates the validity of this observation.[24] Furthermore, the inefficiencies and inequities inherent in across-the-board reductions can be large. Given the high costs of effective long-term mitigation, policymakers should not ignore such inefficiencies. Parson and Zeckhauser, as well as Sebenius, offer several innovative approaches to break this impasse.

INSTEAD OF ESTABLISHING TARGETS FOR NATIONAL EMISSIONS REDUCTION, REGIONAL TARGETS SHOULD BE ESTABLISHED

Under such schemes, each bloc of countries would face a reduction target and then would have to negotiate among the member countries as to how to best allocate responsibility for meeting that target. Such a process would allow for much greater diversity within each bloc. For example, the countries that comprise the European Union have attempted to forge a single unified position for reducing greenhouse gases. While in 1994 there were still differences among the member countries, substantial progress had been made toward an emissions-reduction strategy. It would not be difficult to envision similar discussions within an East Asian trading bloc or a coalition of Eastern European nations.

IN SELECTING A FORMULA FOR DEFINING NATIONAL EMISSIONS ENTITLEMENT, PARTIES SHOULD CONSIDER USING A WEIGHTED AVERAGE OF THREE TO FIVE FACTORS

Parson and Zeckhauser use as an example of a successful formula the International Monetary Fund's (IMF) system of designing quotas, which involves a formula based on national income, exports, imports, reserves, and the variability of exports. This option could be refined further by having two or three different formulae—one for industrialized countries, one for newly industrialized nations, and one for developing countries. Finally, a formula could be devised to create a currency for the trading of emissions reductions over time or across national borders.

NEGOTIATORS SHOULD BE OPEN TO THE POSSIBILITY OF SMALLER GROUPS OF INDUSTRIALIZED STATES THAT ARE INTERESTED IN ANTI–GREENHOUSE GAS MEASURES NEGOTIATING SUBREGIONAL PROTOCOLS

Agreements within subregional groups will likely prove far easier to achieve than trying to structure a consensus among all nations, which always threatens to reawake North-South polemics. For example, the United States might be able to fashion an agreement with the European nations, Japan, and Canada. Once this agreement is in place, the group could enter into bilateral negotiations with developing countries, such as China, Brazil, or India. Bilateral agreements could be tailored to meet the concerns and needs of each nation. As Sebenius points out, such bilateral discussion would allow greater flexibility in the design of incentives than would be permitted within the framework of a larger multinational negotiation. For instance, the package negotiated with China could be substantially different from that discussed with Brazil. For China the emphasis might be on reducing energy subsidies and substituting natural gas for coal, while with Brazil it might prove more fruitful to focus on reforestation programs and energy efficiency investments.

LINKAGES AND SIDE PAYMENTS SHOULD BE ENCOURAGED

If the United States places a high value on reducing greenhouse gas emissions, but another country perceives population growth or soil erosion as a more serious long-range environmental problem, the United States could provide additional resources and funds to help with family planning and improved soil conservation practices. In return, that country could begin to reduce the energy subsidies that promote the use of fossil fuels. In this way, both parties could maximize the benefits of the investment. The United States buys carbon dioxide emissions reduction, which it values, and the recipient nation receives financial aid to meet its environmental concerns, which it values.

Popular opinion: If pressured, China, India, and other developing nations will reduce their greenhouse gas emissions.

Reality: The most difficult challenge in achieving meaningful global reductions in greenhouse gases will be to structure a portfolio of incentives and policies to persuade developing nations to incorporate emissions-reduction strategies into their economic development plans.

Long-term success in the reduction of greenhouse gas concentrations is impossible without the cooperation of the developing world. By the end of the next century, developing countries could account for 60 to 70 percent of the world's greenhouse gas emissions. Therefore, solutions that only involve the developed world will be unsuccessful.

In developed countries, global warming is primarily considered an energy problem. Policy discussion emphasizes energy efficiency, renewable energy, the phasing out of fossil-fueled power plants, and the movement to alternative-fueled vehicles. In developing countries, the issue is perceived differently; the mitigation of global warming is seen as requiring significant alterations in their economic development goals. Why should countries with per capita levels of GNP one-tenth to one-twentieth of those of the United States, Europe, and Japan be asked to forego their development plans to rectify a problem caused by the industrialized world? More starkly, why should a family in Shanghai earning $600 per year sacrifice so that a family in the United States can continue to pay $1.15 per gallon to fuel their two cars and enjoy the benefits of cheap electricity to power their six major appliances in a 2,000-square-foot home?

China is expected to double its coal use between 1990 and 2000.[25] In 1992 several provinces in China experienced economic growth levels of more than 20 percent.[26] A major impediment to maintaining such growth is the lack of adequate public infrastructure, including electrical generating stations. Will China voluntarily agree to forego expanding its infrastructure because the United States and Europe are worried about the impacts of climate change three decades hence?

Inevitably, any workable, long-term reduction program will require large transfers of money and resources from developed to developing nations. In 1990 the United States provided foreign aid equivalent to only 0.28 percent of its GNP, much of it in the form of surplus agriculture, military equipment, and tied export loans (that is, the loans had to be used to purchase goods and services from the United States).[27] In 1986 the World Bank calculated that only 8 percent of the budget of the U.S. Agency for International Development was for developmental aid to low-income countries.[28] To make any difference, the United States would have to dramatically increase its level of foreign aid and channel all of the additional funds to projects that directly or indirectly reduce greenhouse gases. Furthermore, this program could not be a one-time arrangement. The United States would have to transfer this higher level of assistance every year for the foreseeable future, and U.S. industrialized allies would have to increase their assistance proportionately as well.[29]

Given the economic and fiscal problems facing most of the industrialized world during the first half of the 1990s, it is as unreasonable to expect developed countries to provide massive amounts of additional aid in the near future as it is to expect developing nations to forego their economic development plans. Redistributive negotiations, which invariably will have to occur, have not yet been seriously engaged. At the Earth Summit, negotiators deftly avoided language requiring any new real commitments to development assistance beyond helping with the costs of monitoring and reporting. In addition, to insist on an early resolution of how much additional aid is warranted will fan the ideological flames of the North-South debate and make progress toward an agreement much more difficult. Again, incremental and indirect approaches are more likely to be successful than attempts to structure a single solution acceptable to all countries. Developing countries' demands for massive amounts of additional financial aid in the name of global equity may appeal to the populace back home, but such demands will dampen the interest of governments in Europe and the United States (especially in Congress) to help.[30] Several possible approaches are suggested.

INDIRECT APPROACHES SUCH AS TRADEABLE PERMITS OR OFFSETS SHOULD BE PURSUED

Reaching agreement on explicit exchanges of money among nations is always politically difficult. Yet because the costs of reducing greenhouse gases vary so significantly among countries, regulatory mechanisms such as tradeable permits or offsets may provide a vehicle for making such transfers without involving governments, except as sanctioning bodies. As the chapters by Stavins and Hahn and by Parson and Zeckhauser point out, arrangements that allow private or government-owned enterprises to meet their domestic requirements for reductions in greenhouse gas emissions by buying reductions from sources in other countries may have a better chance of garnering political support. The framework convention allows developed countries (Annex 1 parties) to receive credits toward their own emissions objectives by reducing emissions in other countries. This joint implementation concept, while controversial, provides an attractive means to meet the goals of greenhouse gas reductions while increasing investments in recipient countries. Some developing countries, however, fear that industrialized nations will forego their fair share of domestic reductions by buying reductions elsewhere with monies that are currently earmarked to meet existing commitments for financial assistance. Other developing countries, such as Poland and Thailand, have aggressively pursued joint implementation projects.

HOW RESOURCES ARE DEPLOYED IN A DEVELOPING COUNTRY SHOULD BE AS IMPORTANT AS THE TOTAL AMOUNT OF RESOURCES TRANSFERRED

To grow, developing countries will have to make enormous investments in public and industrial infrastructure over the next twenty years. These investment plans may provide a unique opportunity to substitute technologies with low–greenhouse gas emissions for the higher emitting technologies that would otherwise be used. If donor countries can design effective programs of technological cooperation, perhaps developing nations can bypass a generation of technologies with high–greenhouse gas emissions and jump to one of lower emitting technologies.

Norberg-Bohm and Hart stress the need to build the institutional capacity in developing countries to allow for the assimilation and deployment of new technologies into their economic, cultural, and political structures. Coalitions and constituencies for these technologies must be in place; otherwise entrenched interests will oppose their dissemination. For example, if a U.S. company attempts to establish an operation in a developing nation to market environmentally benign products, it may threaten local producers, who will appeal to their allies in government to oppose or impede the dissemination of these technologies on protectionist principles. Admittedly, developing countries have at times embraced new, foreign entrants; nonetheless, the course of least resistance would be to consider partnerships between enterprises in developed and developing nations.

Countries differ dramatically in their receptivity to specific technologies, even when the technologies are designed to meet the perceived needs of these countries. People might be ignorant of the technology's value or unfamiliar with how to use it, and so delay investing in it. Even if they eventually do make the investment, they often underuse the technology. Too often, the selection of the technologies to be transferred reflects the perspectives and values of industrialized countries and not those of the consumers in nonindustrialized nations. The developing world is littered with Western technologies that were inappropriate and thus became relatively useless.[31]

In contrast, experience suggests that a program in which U.S. companies form cooperative relations with local firms to produce and market environmental technologies is much more likely to succeed. If this effort is initiated in tandem with additional U.S. aid to build up the human infrastructure and to establish institutions to support the dissemination of such technologies, a constituency will be in place with a strong self-interest in having the technologies succeed. Cooperative ventures provide the best, and sometimes the only, way to facilitate the dissemination of environmentally superior technologies; they can become win-win situations for all parties involved. Furthermore, such approaches can help establish the institutional infrastructure and human capital to support additional reductions in the future, if such reductions are needed. To persuade the developing countries to make such reductions now, before the political support and the infrastructure are established, would be a prescription for failure.

DESIGNING DOMESTIC INITIATIVES

In designing domestic initiatives, U.S. policymakers should, first and foremost, remember Vernon's admonition: the development of international negotiation strategies and the design of domestic policy initiatives are inextricably linked. To push for protocols and policies at the international level that are unlikely to be acceptable to the U.S. Congress, or the domestic body politic, could result in a serious loss of credibility. Conversely, to design a domestic program that is out of symmetry with the United States's international position could seriously undermine the ability of U.S. negotiators to reach meaningful agreements. A domestic program that exceeds the targets and requirements acceptable to a majority of other nations encourages massive free riding and could result in negligible net reductions in carbon dioxide. Since the climate change negotiations will last several decades, the long-term credibility of U.S. negotiating positions is a significant concern.

In April 1993, President Clinton announced that the United States would stabilize greenhouse gas emissions at 1990 levels by the year 2000, but he was purposely vague on whether this goal would serve as a cap in the years beyond 2000. As mentioned earlier, such a goal, even if reached, is a relatively small step toward the reduction of the growth in greenhouse gas concentrations. Assuming that the U.S. economy continues to grow at current rates, this goal will require ever-increasing reductions over time; reductions after the year 2000 to maintain 1990 emissions levels will be greater than those prior to 2000, since emissions from every new source will have to be offset with reductions elsewhere. To date, the United States has pieced together a collection of voluntary and small initiatives to meet the stabilization goals, but eventually it will have to take actions that will demand greater expenditures of funds, resources, and political capital. If the goal is more ambitious, such as a decrease below 1990 emissions, the costs will be higher and will have to be paid earlier.

Some people argue that since there is so much uncertainty surrounding the science of climate change, the United States should adopt a no-regrets policy. That is, it should only pursue actions that can be justified on grounds other than the reduction of greenhouse gases. Such a policy might include incentives for increased energy efficiency, but only if the costs could be supported by benefits such as cleaner air, greater energy security, or some other indices unrelated to climate change. This type of policy only makes sense if one believes that the probability for global climate change is very low or that the resulting dislocations and harm will be minor. Otherwise, taking some precautions against the possibility of global warming may be a prudent course to take.

There are hundreds of reports, articles, and books that advocate various responses, ranging from massive investments in reforestation or renewable energy to the direct or indirect reduction of carbon dioxide emissions. Most of these proposals fall into five categories: (1) voluntary programs, (2) subsidies and other forms of aid to encourage the development and dissemination of new technologies, (3) command-and-control regulation, (4) taxes, and (5) tradeable permits. Obviously, these categories are not mutually exclusive and often can complement each other.

Stram lists a number of criteria for the evaluation of alternative policy mechanisms. Four of these are of special significance for the design of domestic responses to the threat of global climate change: (1) whether the policy will achieve the stated goal, (2) whether the policy will achieve this goal at the least cost to society, (3) whether the policy will encourage the evolution of new environment-saving technologies or discourage technological innovation and diffusion, and (4) whether the purpose and nature of the policy will be broadly understood and politically acceptable. Other important criteria address topics such as fairness, flexibility, and administrative feasibility; however, the immediate discussion focuses on these four.

If the five policy alternatives are measured against these four criteria, two conclusions become apparent. First, officials will embrace more than one type of response. In fact, over time, the U.S. response will probably include elements of all five policies. Second, the more effective the policy, the more politically difficult it will be to garner support for it. In the remaining pages, I assess these response categories, drawing from each chapter the lessons for the design of a U.S. greenhouse gas reduction strategy. I conclude by examining two specific policy options: carbon dioxide offsets and carbon taxes.

VOLUNTARY ACTIONS

Voluntary actions were a major part of the Bush administration's arsenal of responses to environmental problems. The Green Lights program, which asked corporations to pledge to invest in efficient lighting technologies, and the 33/50 program, which was aimed at convincing 600 companies to voluntarily reduce their 1988 emissions levels of seventeen toxic pollutants 33 percent by 1992 and 50 percent by 1995, are prime examples.[32] The Clinton administration's Climate Change Action Plan contained forty-seven separate initiatives that involved voluntary reductions by industry. For example, one initiative, the Climate Challenge program, called on electric utilities to voluntarily commit to reducing their emissions of greenhouse gases.

Certainly, since such policies minimize the costs to government and are easily accepted by the political establishment, opposition will be negligible. The problem is that voluntary programs are not apt to achieve the emissions reductions needed to stabilize emissions at 1990 levels. If the reductions that would have occurred without the existence of these programs are subtracted from overall reductions, the savings from such programs fall short of the president's goal. The ineffectiveness of voluntary measures stems from the fact that all of the benefits do not accrue to the parties making the investments. Reducing greenhouse gas concentrations is a global problem, and net reductions from any one source benefit everyone on the planet. That is, the benefits are dispersed, while the costs remain localized. A utility company would be hard-pressed to convince its local regulators that electric rates should increase $20–30 million to cover the cost of planting trees in another region or the cost of converting from a cheap high–carbon dioxide emitting fuel to a much more expensive low–carbon dioxide emitting fuel.

SUBSIDIES TO SPUR NEW TECHNOLOGIES

Americans are fascinated by technology. Whether it is the Manhattan Project or putting a man on the moon, Americans often assume that there is no problem too big or too difficult that it cannot be solved by the right technology. The threat of global climate change is no different. Some experts espouse massive development of solar photovoltaic cells.[33] Others support energy efficiency, while still others advocate a new generation of nuclear plants with an eventual transition to a hydrogen economy.[34] (The Clinton administration's Climate Change Action Plan calls for several public-private partnerships to develop advanced energy efficiency technologies, alternatives to the internal combustion engine, and a wide variety of renewable options.) Proponents of adaptation list technologies that are or will soon be affordable, such as new drought-resistant crops or better sea wall technologies. Other experts espouse investing in geo-engineering options, such as seeding the oceans with iron to stimulate carbon dioxide–absorbing phytoplankton, screening sunlight by placing mirrors in space, or shooting dust particles into the stratosphere.[35]

Few would argue that technology will not play a significant role in meeting the threat of global climate change. The issue is how best to spur the development and dissemination of new technologies. One option is to subsidize those technologies that government experts believe hold the most promise. But many have little faith in government's ability to make the correct selections. Stram reviews the lack of success of past government efforts, such as President Carter's Synfuel Corporation, which spent billions of dollars developing technologies for which there was no market.

Overreliance on government subsidies invariably presents three problems. First, vested interests in existing energy technologies assert strong political pressure on the executive branch and Congress to subsidize technologies that are an extension of the existing energy mix. For example, in the 1970s and 1980s, coal and nuclear power received far more support than energy efficiency programs and renewable energy, in part because the former industries were able to marshal the support of hundreds of established businesses, unions, government laboratories, and other organizations economically linked to them. Energy conservation and renewable energy were not as established, and their advocates had more difficulty in garnering political support and funding for research and development. Second, even the most knowledgeable experts cannot predict which technologies will be successful in terms of marketability and effectiveness ten years later. Large subsidies will inevitably go to technologies that will not live up to their promise, and as a result, they will not be cost-effective. Finally, it is extremely doubtful that the federal budget will permit subsidies large enough to spur investment in new technologies sufficient to meet emissions-reduction goals.

COMMAND-AND-CONTROL REGULATION

Command-and-control regulation, such as establishing emissions levels for all sources of a certain size (emissions standards) or requiring that all sources of a certain type install a particular piece of equipment (performance technology standards), has historically been the primary instrument used to meet U.S. environmental goals.

A set of command-and-control regulations could be designed to achieve any particular greenhouse gas reduction goal. However, it almost certainly would not be cost-effective. Command-and-control regulations require all firms to meet a preset standard, regardless of cost; that is, businesses are all treated alike, whether it costs them a lot or only a little to comply. As Hahn and Stavins point out, the cost of controlling a given pollutant could vary by a factor of one hundred or more among sources. This variability is ignored by command-and-control approaches. On the other hand, such approaches have a proven track record of garnering political support in the U.S. Congress, and at least in the case of technology-based standards, the monitoring and enforcement costs are comparatively low.

Most studies indicate that even an efficient program to reduce carbon dioxide emissions will require an unprecedented level of investment relative to all other environmental programs. It will be difficult to obtain public support for such expenditures under the best of circumstances. Therefore, additional costs in the form of wasteful use of resources resulting from inefficient command-and-control regulations are likely to meet with political resistance. Some regulation will be inevitable if the United States decides to seriously reduce carbon dioxide emissions, but this regulation needs to be part of a much larger program.

Theoretically, while command-and-control regulation can push some technologies into the marketplace and can require the purchase of others, it can also lock in the existing array of technologies by removing the incentive for a company to exceed its emission targets. Furthermore, regulators rarely require the use of a technology that is not already available and proven. As Hahn and Stavins point out, under command-and-control regulation, money that could be invested in technology development too often gets diverted to lobbying and litigation.

ENERGY TAXES

The basic principle behind energy taxes is that by making fossil fuels more expensive, consumers will use less, and therefore carbon dioxide emissions will be reduced. The most effective tax to reduce carbon dioxide would be a carbon tax, since it is targeted at the primary source of the problem. But as Jorgenson and Wilcoxen argue, such a tax would fall disproportionately on the coal industry and, to a lesser extent, on industries that use coal. Other taxes that fall within the definition of an energy tax are Btu and gasoline levies.

Energy taxes could achieve President Clinton's reduction target and would do so at costs significantly lower than equivalent command-and-control regulatory programs, since taxes would provide the most effective incentives to firms to limit their emissions. Taxes would also provide polluters with a systematic incentive to select the most effective methods. Furthermore, taxes provide clear and strong incentives to encourage the development and dissemination of new technologies. Their principal drawback is that taxes are a political lightning rod, as they attract the opposition of vested interests.

EMISSIONS TRADING

A second type of market incentive program is emissions trading, sometimes called tradeable permits. Under a tradeable permit scheme, the government specifies in advance the target level of emissions (that is, the cap) and then allocates emissions rights or allowances to emitting firms in the form of permits. Some firms will find it relatively inexpensive to reduce their emissions in excess of the number of permits allocated to them, while others will find it much more difficult to reach the same level of reduction. Under a tradeable permit scheme, the former can sell permits to the latter, thus lowering the cost of meeting the emissions-reduction target from the cost of complying with traditional command-and-control regulation.

A tradeable permit regime could take many forms. It could cover carbon dioxide alone or all greenhouse gases. It could include only sources or both sources and sinks. It could allow trading only within a nation's borders, or it could be designed to facilitate trading across countries. Tradeable permit schemes have many of the same attributes as carbon taxes: they can achieve greenhouse gas reduction goals, and do so in a cost-effective manner, and they can stimulate the transfer of new technologies. Tradeable permit schemes are more politically controversial than voluntary programs, technology subsidies, or command-and-control regulation, but much less so than energy taxes.

Critics of tradeable permits, energy taxes, or emissions fees argue that some studies have ignored obstacles such as transition problems, transaction costs, administrative costs, and political feasibility and that the actual savings will be far less than those predicted by theorists. They are correct. But the issue is whether these types of market incentive initiatives are more cost-effective than the alternatives, not whether they realize some theoretical optimum.

What, if any, are the strategic differences between tradeable permits and energy taxes, and, more specifically, carbon taxes? The chapters by Stram and by Hahn and Stavins discuss several differences. First, while both options impose costs on industry and consumers, carbon taxes make these costs much more politically visible. Stram argues that this visibility is an advantage: since consumers will know precisely what it will cost them to reduce carbon dioxide, they will be forced to choose how much insurance against global climate change they really want to buy. That is, a carbon tax plays a valuable role in educating the public. Hahn and Stavins argue the opposite, pointing out that such visibility may make it much harder to obtain political support for carbon taxes, especially compared with other market mechanisms, such as tradeable permits.

Second, international tradeable permit schemes can provide a vehicle for transferring technology and aid to the developing world. A company in the United States could purchase reductions in Brazil at one-quarter the cost of reducing its own carbon dioxide emissions. Theoretically, the United States should be neutral concerning whether the reduction takes place in Brazil or within the United States, since the benefits, as measured by lower carbon dioxide concentrations, are the same. Tradeable permits provide an indirect mechanism for funneling financial resources to the developing world at a time in which it will be difficult to obtain increases in government-to-government foreign aid.

As suggested by Stram, the United States could funnel a portion of the revenues from a carbon tax into an international fund and achieve some of the same benefits. Congress, however, would have to agree to establish such a fund, appropriate the money, and resist the temptation not to divert existing foreign support. Furthermore, it is hard to imagine that Congress would not require recipient nations to guarantee that the funds were used to provide incremental environmental progress. Designing the mechanisms for meeting this requirement would challenge the ingenuity and creativity of both U.S. agencies and officials in recipient countries. Perhaps Congress will demonstrate such a commitment, but certainly an international fund would be more difficult to achieve and maintain—at least on the domestic political front—than would a tradeable permit scheme.

A full-blown international tradeable permit scheme will be difficult to design and administer given the number of parties that would have to agree to a myriad of details (monitoring protocols, banking provisions, initial allocations of permits, etc.). The United States is still working to fine-tune its sulfur dioxide allowance scheme to reduce acid rain; if the

U.S. government cannot get several hundred domestic electric utilities, most of which are regulated monopolies, to participate, how will the UN or some other international body be able to implement an international permit regime involving thousands of sources? Furthermore, what guarantee would there be that reductions in Brazil would actually take place and would not be reversed if and when the government changed? If such programs were established on a large scale, an elaborate monitoring system would have to be agreed on and put into place, as would international institutions to oversee and enforce the trading regime. Also, some U.S. environmentalists would, at least initially, see such a program as allowing polluting industries to avoid their obligations to reduce their emissions.

As mentioned earlier, the United States will go through several cycles of policy development; options not embraced in one cycle could play a significant role in the next. Most of the chapters in this book recommend starting with responses that are designed to maximize the number of supporters and minimize the number of blocking coalitions. Small, precedent-setting steps will be more effective than large, controversial ones.

THE POTENTIAL OF CARBON DIOXIDE OFFSETS

Initially, the United States might support a program promoting carbon dioxide offsets, which is equivalent to a reduced version of the tradeable permit concept—emissions rights equal to current total emissions would be allocated to existing sources in proportion to their present emissions levels or another equivalent benchmark. New sources would have to buy permits from an existing source, which would then have to reduce its emissions by an equivalent amount. In the case of carbon dioxide, one proposal would have the new facility fund reforestation projects or pay another source to reduce carbon dioxide emissions by investing in energy efficiency, fuel switching, or new processes. An offset program could be limited to transactions within the United States or could be extended to allow transactions with firms in other countries where costs might be considerably lower. The major problem with an international program is that the institutions needed to monitor and enforce emissions trades are not in place.[36] The Framework Convention encouraged developed countries to pursue joint implementation projects with each other. While there is resistance to the concept in some developing countries, others have expressed an active interest in being included in a pilot program. These initial efforts will be of enormous assistance in resolving the outstanding problems. To a large extent, joint implementation provides a foundation from which the international community and, more specifically, the United States can erect a workable offset program.

The advantage of establishing an offset program is that it would allow all of the participating parties—private and public, U.S. and foreign—to familiarize themselves with many of the implementation issues that would accompany a full-blown tradeable permit program. Offset programs would also establish a precedent for the use of market incentives, which could be expanded at a later date. It would be difficult to imagine a scenario in which nations went from having no program to implementing a comprehensive international regime of tradeable permits. Even though the United States may have some familiarity with the domestic use of similar schemes, Europe and Japan are only beginning to assess their feasibility and the developing world has no experience with them. A carefully targeted offset program could represent an effective first step. If such a program proved to be administratively workable and could be expanded into the international arena, it could become an effective vehicle for transferring financial assistance, technologies, training, and resources to the developing world.

The major difficulty with any tradeable permit scheme is the establishment of administrative and enforcement mechanisms. The greater the number of sources to be covered, the more unmanageable the program becomes. Therefore, it is essential to start small. Starting with joint implementation projects, moving to an offset program, and then eventually, if the need arises, adopting an international tradeable permit program is a progression which, over ten to fifteen years, might be workable.

A mechanism would have to be devised to limit the potential of existing sources to use the program to assert market power. Initially, the program would have to be limited in scope, but the goal would be to reach bilateral, and perhaps multilateral, agreements with a select group of countries to establish the institutions to monitor cross-country trades.

The purpose is not necessarily to use this program to substantially reduce greenhouse gas emissions, but rather to put in place the institutions and expertise to administer such programs. Voluntary programs, subsidies, or even most command-and-control programs do not have the capacity or flexibility to be rapidly expanded (or, if the evidence warrants, decreased). Tradeable permits and taxes do, and this flexibility can be a major advantage when a country is attempting to respond to a problem characterized by great scientific uncertainty, multiple program and policy cycles, and fluctuating political dynamics.

THE POTENTIAL OF CARBON TAXES

What about a carbon tax? Jorgenson and Wilcoxen point out three important features of such a tax. First, the level at which a carbon tax needs to be set in order to hold U.S. carbon dioxide emissions to a specific target will vary from year to year. For example, if the United States chose to stabilize carbon dioxide emissions at 1990 levels, Jorgenson and Wilcoxen suggest that the tax would be about $6 per ton in 2000, about $14 per ton in 2010 and about $23 per ton in 2020. The United States would have twenty-seven years to adjust before it had to pay $23.[37] These numbers are a product of a model, and models can be imprecise, but the message is that a carbon tax could be effectively phased in over several decades.

Second, the impact of the tax on the U.S. economy depends on what the government decides to do with the revenue. Jorgenson and Wilcoxen assert that if the revenue were used to reduce distortionary taxes elsewhere, the impact of the tax on U.S. GNP would be substantially lessened. In fact, under certain carbon tax-revenue recycling schemes, GNP actually increases.

Third, no matter how they are designed, carbon taxes will have a devastating impact on the coal industry. Jorgenson and Wilcoxen predict that stabilization reached solely by a carbon tax would reduce coal output by almost 30 percent by 2020 and would increase prices by 46 percent. If the nation used the same policy to achieve a 20-percent reduction, coal prices would rise 150 percent and output would drop by 55 percent. The impact on the coal industry is five times larger than that for the next most affected industry—the electric utilities. Coal-producing states will fight such initiatives, and without determined support from other interests, they will have the power to block an effective carbon tax. The Clinton administration's experience with the Btu tax in 1993 illustrates the political gauntlet that even a proposal for a relatively small energy tax must endure.

If the United States decides it must reduce carbon dioxide emissions, the substantive arguments for a carbon tax are compelling. A carbon tax is one of the few cost-effective policies that could stimulate measurable decreases in carbon dioxide emissions. It will remain a major focus of the environment and energy debate in Europe and Japan. However, proponents for a U.S. carbon tax must first tackle the United States's visceral political aversion to energy taxes, and the only way to face this problem successfully is to provide tangible benefits in return for the higher taxes. Energy taxes are not politically popular as a means to reduce the deficit, but if, as Jorgenson and Wilcoxen suggest, they are sold as part of a larger package to reduce distortionary taxes elsewhere in the economy and if the beneficiaries of this reform are clearly identifiable, the public's reaction might be different. There are goals and priorities for which the American public will be willing to pay higher energy prices. The benefits cannot, however, be abstract, diffuse, indirect, or difficult to measure; a majority of Americans must be able to understand clearly that if they are paying 50 cents more for a gallon of gasoline, they are getting something of equal or greater value in return. Protection from future global climate change does not currently meet this test. But nowhere in this book do we argue that all of the revenues should primarily go to mitigating the threat of global warming. In fact, it would probably be politically impossible to sell an energy tax on the grounds that it is needed to reduce the threat of climate change. Rather, the tax must be shown to move the tax code away from penalizing the efforts of labor and investment, which most people want to see supported, to discouraging the use of fossil fuels, which people might be willing to use less of (although there is no evidence that Americans have reached this conclusion).

The visceral aversion to energy taxes, such as a carbon tax, is a political barrier that must be breached before the United States will ever view such taxes from the same perspective as the Europeans and the Japanese. It would be a serious mistake for policymakers either to ignore the existence of this barrier or to view it as an insurmountable obstacle. If the United States is to build an institutional foundation capable of responding to the problem of global climate change, its policy arsenal needs to include the ability to establish, expand, and use both tradeable permits and carbon taxes. At the moment, the United States can implement neither. An ambitious goal would be to change this situation by the turn of the century.

SUMMARY

Should the United States respond to the possible threat of global warming, and how large and far-reaching should that response be? It is probably fair to say that most of the authors in this volume agree that some mitigation response is justifiable, and all understand that determining the magnitude of the response entails a political decision that will be made through the democratic process. In fact, as mentioned earlier, we believe that decisions will be remade several times during the next two decades, as continuous international negotiations respond to new scientific and political information.

The chapters that follow offer guidance and broad principles that can be helpful in designing strategies and responses to the global climate problem. If three themes can be drawn from these chapters, they are (1) that policy initiatives and programs should be advanced in small steps, each building a foundation for the next; (2) that the flexibility to respond to scientific, economic, and political change will be essential; and (3) that decision-making processes must be designed that clearly link international and domestic considerations with their respective stakeholders.

NOTES

1. Phillip Williamson, *Global Change: Reducing Uncertainties* (Stockholm: International Geosphere-Biosphere Programme, June 1992), 4.
2. J. T. Houghton, G. J. Jenkins, and J. J. Ephraums, eds., *Climate Change: The IPCC Scientific Assessment: Report Prepared for IPCC by Working Group I* (New York: Press Syndicate of the University of Cambridge, 1990).
3. Richard Lindzen, "Absence of Scientific Basis," and Patrick J. Michaels, "Benign Greenhouse," *Research and Exploration* [a scholarly publication of the National Geographic Society] 9, no. 2 (spring 1993): 191–200 and 222–233, respectively.
4. Scientific circles disagree as to whether CFCs, which are also a major contributor to depletion of the stratospheric ozone layer, have a net warming or cooling effect. At the time this chapter was written, the IPCC was still recommending their inclusion as a major greenhouse gas, albeit the United States had opted not to count reductions of CFCs in meeting its goal of stabilizing greenhouse gases at 1990 levels by the year 2000.
5. Houghton et al., *Climate Change*.
6. Ibid., 7. This report points out that the way in which carbon dioxide is absorbed by the oceans and biosphere is not simple, and a single value cannot be given.
7. Daniel A. Lashof and Dennis A. Tirpak, eds., executive summary to *Policy Options for Stabilizing Global Climate* (Washington, DC: U.S. EPA, Office of Policy, Planning, and Evaluation, December 1990), 9.
8. William D. Nordhaus, "An Optimal Transition Path for Controlling Greenhouse Gases," *Science* 258, no. 5086 (20 November 1992): 1317.
9. Ibid., 1318.
10. Lonnie Thompson and Ellen Mosely-Thompson, "Evidence of Abrupt Climate Change During the Last 150,000 Years Recorded in Ice

Cores from the Tropical Quelccaya Ice Caps," in *Abrupt Climate Change: Evidence and Implications*, eds. W. H. Berger and L. D. Labeyrie (The Netherlands: Reidl, 1987); and Greenland Ice-Core Project Members (GRIP), "Climate Instability during the Last Interglacial Period Recorded in the GRIP Ice Core," *Nature* 364, no. 6434 (15 July 1993): 203–207.

11. Harvey Brooks, "The Typology of Surprises in Technology, Institutions, and Development," in *Sustainable Development of the Biosphere*, eds. W. C. Clark and R. E. Munn (Cambridge: Cambridge University Press, 1986), 325–350.

12. International Energy Agency, *Climate Change Policy Initiatives*, vol. 1 (Paris: OECD/IEA, 1994), 15.

13. The further one goes into the future, the less the difference in costs between reaching various reduction targets. That is, cost curves begin to merge as they approach the twenty-second century. For example, the cost of stabilizing emissions at 1990 levels is relatively low from 1993 to the year 2000. It gets more expensive between 2000 and 2010 as economic growth creates more emissions sources, and even more expensive between 2010 and 2020. If this analysis is carried out to 2100, the cost of stabilizing at 1990 levels becomes very high in the outer years, and the initial costs of a 20-percent reduction become less important.

14. Responding to concern over dependence on foreign oil, President Ford used presidential powers in 1975 to raise the duty on imported oil. His Energy Independence Act of 1975, sent to Congress on 1 January 1975, proposed a domestic windfall profits tax, an excise tax on oil and natural gas, and various tax credits for reducing energy consumption. President Carter's Moral Equivalent of War speech (18 April 1977) announced his National Energy Program, which included a gasoline tax, a wellhead tax, and a gas-guzzler excise tax on cars that did not meet mileage standards. President Bush, in a deficit-reduction compromise for the FY1991 budget (announced 30 September 1990), agreed to an increased gasoline tax. President Clinton's original proposal for a Btu tax was changed to a four- to five-cent increase in the gasoline tax for the FY1994 budget.

15. At the time this book was written, there was renewed debate among the relevant EU ministers about this tax. Germany and the Benelux countries advocated the imposition of such a tax as a precondition of ratifying the framework convention, while Great Britain, Spain, Portugal, and Greece resisted on the grounds that their economies were much less developed than those of other EU members. Great Britain

opposed both carbon and energy taxes, but did commit to changes in their VAT tax structure, which would include domestic fuel prices. As of January 1995, no agreement had been reached on a European energy or carbon tax. See "Current Reports," *BNA's International Environmental Reporter,* various issues, 1993–1995.

16. UN General Assembly, Intergovernmental Negotiating Committee for a Framework Convention on Climate Change, 1992, "Framework Convention on Climate Change," UN doc. A/AC.237/18 (Part II)/Add.1, 15 May. Each country listed in Annex 1 to the convention is committed to "communicate" to the Conference of Parties detailed information on its climate change policies and their effect on greenhouse gas emissions. Countries which ratified that convention by 21 March 1994 are scheduled to submit their communications by 21 September 1994.

17. Chapters 35 and 40 of UNCED's Agenda 21, which sets forth the policies and programs to realize the goal of global sustainable development, call for the establishment of unified formats and systems for collecting and publishing data and information.

18. There are several examples of this process of using reports to stimulate responses from interest groups and experts. Probably the most well-known are those in the human rights realm. In the area of the environment, the Europeans require that reports on actions taken to reduce sulfur dioxides be submitted to a central secretariat and publicly scrutinized. There are some who believe that this process was effective in pushing Great Britain to reduce such emissions. See Marc A. Levy, "European Acid Rain: The Power of Totebond Diplomacy," in *Institutions for the Earth: Sources of Effective International Environmental Protection,* eds. Peter M. Haas, Robert O. Keohane, and Marc A. Levy (Cambridge: MIT Press, 1993), 123–125.

19. B. Bolin, B. Doos, J. Jager, and R. A. Warrick, eds., *The Greenhouse Effect, Climate Change, and Ecosystems: SCOPE* 29 (Chichester, England: Wiley, 1986).

20. Peter M. Haas, "Introduction: Epistemic Communities and International Policy Coordination" and "Banning Chlorofluorocarbons: Epistemic Community Efforts to Protect Stratospheric Ozone," and Emmanuel Adler, "Conclusion: Epistemic Communities, World Order, and the Creation of a Reflective Research Program," in *International Organization* [Special Issue: Knowledge, Power, and International Policy Coordination, ed. Peter Haas] 46, no. 1 (1992).

21. Robert Livernash, "The Growing Influence of NGOs in the Developing World," *Environment* 34, no. 5 (June 1992): 12.

22. Richard E. Benedick, *Ozone Diplomacy* (Cambridge: Harvard University Press, 1991).

23. In August 1994, the INC considered a proposal whereby the central staff would prepare a summary document synthesizing the contents of the National Communications (action plans), but actual reviews would be conducted by panels appointed and staffed by the participating countries' governments.

24. By the summer of 1994, four of the OECD countries (Spain, Portugal, Greece, and Ireland) had been willing only to commit to limiting the growth in greenhouse gas emissions, and one (Turkey) had not signed the convention. Five countries had chosen to reduce net emissions, while eleven had pledged to meet gross emissions-reduction targets. Two countries, Japan and France, had committed to stabilize per capita emissions.

25. Anthony Cortese, "Clearing the Air," *Environmental Science and Technology* 24, no. 4 (1990): 444; or International Energy Agency, *Global Energy: The Changing Outlook* (Paris: OECD/IEA, 1992).

26. *China Statistics Monthly*, University of Illinois at Chicago, China Statistical Information and Consultancy Service Centre, Beijing, various editions, 1992.

27. *Statistical Abstract of the United States*, 1992 (Washington, DC: U.S. Department of Commerce, 1992).

28. *World Development Report 1990: Poverty* (Washington, DC: World Bank, 1990), 128.

29. Patrick McCully warned that massive aid programs may do more harm than good, and cautioned against their use without significant economic and political restructuring in both the developed and developing nations (McCully, "The Case against Climate Aid," *The Ecologist* 21, no. 6 [November–December 1991]: 244–251).

30. The Centre for Science and Environment in India recommended that high–greenhouse gas emitters pay approximately $100 billion per year to the rest of the world. See Anil Agarwal and Sunita Narain, *Global Warming in an Unequal World: A Case of Environmental Colonialism* (New Delhi: Centre for Science and Environment, January 1991), 20.

31. McCully, "The Case against Climate Aid," 248.

32. The EPA's Green Lights program, started in 1990, provides funding and information for community energy conservation programs. The 33/50 program was announced by William Reilly as a result of the EPA's Science Advisory Board report *Reducing Risk: Setting Priorities and Strategies for Environmental Protection. Appendix C: The Report of the*

Strategic Options Subcommittee, SAB-EC-90-021 (Washington, DC: U.S. EPA, September 1990).

33. Michael Oppenheimer and Robert H. Boyle, *Dead Heat: The Race against the Greenhouse Effect* (New York: Basic Books, 1990).

34. Most visions of a hydrogen fuel-based economy involve the use of nuclear power or solar photovoltaics to break water into hydrogen and oxygen. The resulting hydrogen can be used as a storable and transportable fuel, which yields water (and small quantities of nitrogen oxides) as a by-product when it is burned. Technical problems remaining to be overcome include the development of cheaper photovoltaics and of safer nuclear technologies.

35. *Policy Implications of Greenhouse Warming—Synthesis Panel* (Washington, DC: National Academy Press, 1991), 113.

36. Gary Marchant, "Freezing Carbon Dioxide (CO_2) Emissions: An Offset Policy for Slowing Global Warming," discussion paper no. 91-2, Center for Science and International Affairs, Harvard University, March 1991, 16.

37. According to this model, a $22.71 per ton of carbon tax would increase coal prices by 49.99 percent, electricity prices by 6.60 percent, and oil prices by 4.45 percent.

2 Overcoming Obstacles to a Successful Climate Convention

■ **JAMES K. SEBENIUS**

Harvard Business School
Harvard University

One key legacy of the mammoth Earth Summit in June 1992 is a framework convention intended to control climate change. Signed by 153 nations following sixteen months of negotiation within the INC, the climate convention will enter into force ninety days after the fiftieth ratification is received. Many environmentally concerned citizens regard the set of principles embodied in this document as essentially meaningless since the agreement lacks any specific commitments or timetables for the reduction of greenhouse gas emissions or any concrete financial commitments from developed countries to enable developing countries to pursue more greenhouse-friendly development strategies. Other advocates are more circumspect about the result, believing it to be the essential, if cautious, first step toward controlling human actions that may lead to damaging climate change. Still others find grounds for optimism in the unprecedented scale of governmental and nongovernmental participation in the Earth Summit and in the white-hot glare of publicity generated worldwide on behalf of environmental issues. Whatever the ultimate verdict, the present climate convention is likely to be followed by years of on-and-off international negotiations over more specific strategies for controlling global warming.

This chapter offers an assessment of the climate talks thus far and asks how best to move them forward. In so doing, it does not seek to evaluate or resolve the considerable scientific and economic uncertainties that surround the global warming issue, nor does it analyze the merits of the many proposed policy responses. In particular, it does not weigh the option of no further international action on the climate issue. Instead, as a reasonable (but contested) assumption for purposes of analysis, this chapter uncritically maintains that the prospect of a serious climate problem exists and adopts the perspective of a greenhouse control advocate who seeks a more stringent type of coordinated international action, but who is agnostic as to its particular form (for example, targets, carbon taxes, or tradeable permits).

In addressing this question, I implicitly draw on concepts from the emerging prescriptive field of negotiation analysis, an approach with roots in game theory, decision analysis, and social psychology.[1] In particular, the analysis is organized around the concept of building and maintaining a meaningful winning coalition of countries that take concrete measures to curb climate change. In a parliamentary context, a winning coalition is a group sufficient to enact legislation. Here, the term is used in a more subjective and expansive sense, defined with respect to the goal of a particular greenhouse gas control advocate. In particular, a winning coalition consists of a set of countries whose actions to control climate change are sufficient to meet that goal. (Different goals thus imply different winning coalitions; this chapter's analysis is framed to apply readily to different goals.) To an advocate, the opposite of a winning coalition is a blocking coalition, or a group able to prevent—passively or actively—a winning coalition from emerging and taking effective greenhouse control actions over time.

There is a temptation among some observers to attribute the lack of progress in the INC talks to the United States and, in particular, to the notable opposition of powerful Bush administration figures (such as former Chief of Staff John Sununu, Budget Director Richard Darman, and Chairman of the Council of Economic Advisers Michael Boskin). It follows, therefore, that a more environmentally sensitive and committed Clinton-Gore administration should unblock progress. This chapter argues that such a view, though not incorrect, is too narrow. There is a much broader range of scientific, economic, and ideological barriers to meaningful and sustainable climate agreement. As such, the present analysis first describes these barriers and then concentrates on how potential and actual blocking coalitions can be prevented from forming, can be acceptably accommodated, or can otherwise be neutralized.[2]

Given the prominence and publicity surrounding international conferences, getting stronger and more specific treaty provisions (that is, protocols) negotiated, signed, and ratified by a suitable number of countries may seem like the obvious route to achieve a meaningful winning coalition. Yet, to a greenhouse control advocate, *the success of a negotiation should not be measured by paper mandates and signatures affixed, but instead by the tangible results it stimulates, directly or indirectly.* These results could, for example, include increased energy efficiency that would not otherwise have occurred, slowed deforestation and new programs of afforestation, and changes from relatively carbon-intensive fuels such as coal to cleaner ones such as natural gas. By analogy, a committed Europhile may have been thrilled at the sight of heads of state signing the Maastricht treaty, which promises closer monetary, economic, and foreign policy integration among the European Community member states. However, if these leaders prove to have been too far ahead of their national constituents— who, like the Danish people, end up voting against ratification—the negotiated instrument may even set back the cause of European unity. If this scenario unfolds, some advocates may wish that there had been greater efforts at consensus building and a less ambitious treaty that might have proved to be a more solid building block. In short, while an advocate's natural focus may be on formal negotiations and far-reaching treaties, these should be viewed as the means to an end rather than as ends in themselves.

A greenhouse control advocate faces an analogous fundamental strategic choice. One option would be to devote primary energies to exploring which diplomatic and procedural devices are most likely to achieve a negotiated instrument that directly mandates strict adherence to greenhouse-friendly policies—a kind of global environmental protection agency with powerful enforcement capabilities. Another approach would be to think about how the negotiations themselves can be used indirectly to achieve their goals by acting as instruments to strengthen the broad and deep consensus needed to overcome the formidable obstacles to action. For example, the negotiating process can be regarded less as a device whose sole purpose is to reach formal binding agreements and more as a means of generating publicity, raising awareness, helping to mobilize sympathetic environmental opinion worldwide, forging coalitions across national boundaries, and providing support for broad-based scientific and policy research to resolve troublesome questions about climate change. Deciding which of these approaches, or what combination of them, holds the most long-term promise for crafting a winning climate coalition depends on the nature and extent of the obstacles to be overcome. It is to an examination of those obstacles—and the actual and potential blocking coalitions to which they give rise—that this chapter turns. First, however, it is useful to place the climate negotiations and the barriers they face in a historical perspective.

THE DIFFICULTY OF ATTAINING AGREEMENT TO CONTROL CLIMATE CHANGE

The widely accepted goal for climate change negotiations has been to generate a general framework convention, perhaps together with one or more protocols on specific subjects. A framework convention was indeed signed in Rio de Janeiro, though specific control protocols were not.[3] Of special disappointment to many greenhouse control advocates worldwide was the virtually single-handed opposition of the United States—among industrialized countries at least—to the adoption of binding targets and timetables for greenhouse gas stabilization in the convention. (Before the 1992 Earth Summit, the nations of the EC and the European Free Trade Association, along with Japan, Australia, Canada, and others, adopted greenhouse gas stabilization or reduction targets.[4] On 21 April 1993, President Clinton committed the United States to return its greenhouse gas emissions to 1990 levels by the year 2000.[5]) In part, this step-by-step, framework-protocol approach was a reaction against the years of negotiating the detailed and comprehensive Law of the Sea (LOS) treaty, which, ultimately, was rejected by the United States and opposed by a few other key powers. In part, the current approach to climate negotiations seeks to build on the perceived success of an analogous process that led to widely accepted control measures for CFCs, an ozone-depleting class of chemicals that also plays some role in greenhouse warming. While a large number of other international negotiations have influenced the dominant course of climate change negotiations and contain useful insights, both the LOS and the CFC negotiations concerned global resources (like the atmosphere). They embody valuable lessons in themselves and serve as especially salient examples for many informed observers.[6]

NEGOTIATIONS OVER OZONE AND OCEANS

The Third United Nations Convention of the Law of the Sea, launched by the General Assembly in 1970, led in 1982 to a comprehensive treaty signed by 159 states (and other authorized parties) that formally enters into force once the sixtieth instrument of ratification is deposited; the required number was reached in 1994.[7] On the positive side, against the predictions of many knowledgeable observers, a broadly acceptable LOS Convention—a "constitution for the oceans"—did result from this mammoth effort despite technical complexity, uncertainty, and ideological division. The negotiation process and the LOS treaty have reduced much of the ocean conflict that was burgeoning at the outset of the negotiations. Given these factors—and the fact that the atmosphere, like the oceans, is a global resource—there were calls from some quarters for a loosely analogous, comprehensive Law of the Atmosphere to address global warming.

By contrast, many view the LOS model as precisely the wrong way to negotiate a convention. The process was conducted at a level of detail that arguably should have been unthinkable in a treaty framework; moreover, twenty years after its inception, the result barely on the threshold of entering into force. In the views of skeptics, the result of this unwieldy process, especially with respect to deep seabed resources, was unworkable, a dangerous precedent, and counter to Western interests. Just as the United States rejected this flawed treaty, goes this line of argument, so should it reject any analogous process or result on climate change.

In parallel with the later stages of the LOS talks, another relevant set of negotiations got under way.[8] In 1977 the United Nations Environment Programme (UNEP) and other UN agencies drew up an Action Plan to Protect Stratospheric Ozone that strengthened international efforts at research, monitoring, and assessment. Under the auspices of the UNEP, a working group was established in May 1981 to try to come up with a global agreement (a framework convention) to protect the ozone layer from CFCs. After seven rounds of negotiations, the compromise Vienna Convention for the Protection of the Ozone Layer was signed in March 1985 by twenty countries and the EC. The Vienna Convention created a framework for international cooperation on research, monitoring, and exchange of information and provided procedures for developing protocols containing specific control measures. In 1987 twenty-four countries signed the Montreal Protocol on Substances that Deplete the Ozone Layer, which calls for the consumption of most CFCs to be cut by 50 percent by 1999. By mid-1990, over sixty countries had ratified the Montreal Protocol or announced their date of ratification. This list included key developed countries, including the United States, the former Soviet Union, Japan, and the EC countries. However, relatively few developing countries had ratified the Montreal Protocol; holdouts included potentially major CFC producers, such as India, China, and Brazil. In a June 1990 London meeting, following some North-South pyrotechnics, ninety-three nations—including some vocal holdouts from developing countries such as India—signed a much strengthened CFC convention that would virtually ban CFC production and use by the year 2000. The new agreement also promised substantial financial and technical assistance to the developing world.

In direct contrast to the blunt U.S. rejection of the LOS treaty, President Reagan described the 1987 Montreal accord as "the result of an extraordinary process . . . of international diplomacy . . . a monumental achievement." In assessing the relevance of this approach for climate change negotiations, some of those involved with the CFC process—after noting that the complexity of climate issues makes it "impossible to deal with everything at once"—recommended disaggregating the problem and following a step-by-step framework-protocol process modeled after the CFC experience. Subsequent official action by both developed and developing countries endorsed the framework-protocol approach.

POLICY RESULTS FROM CLIMATE CHANGE NEGOTIATIONS COMPARED WITH THOSE OVER CFCs OR THE LOS

To place the climate change negotiations in perspective, especially with respect to those over the oceans and the ozone layer, it is important to understand the nature of the issue. Consider four complementary dimensions that can be used to understand the sources of greenhouse gas emissions:

- In conventional scenarios, slightly less than half of the expected warming from emissions during the 1980s, for example, came from energy-related activities (coal, petroleum, and natural gas used in industry, home heating, transportation, etc.). Nonenergy industrial activities delivered about a quarter, and land use activities (deforestation, rice cultivation, fertilization, etc.) caused the rest.

- About 55 percent of the expected contribution to warming from emissions during this period is due to carbon dioxide, with CFCs (24 percent, likely much less depending on the effects of the Montreal Protocol and of findings that the effective warming contribution of CFCs may be small), methane (15 percent), and nitrous oxides delivering the rest.

- About half of the expected warming will reflect global population growth and about half will reflect growth in per capita demand.

- About 40 percent of the expected warming now comes from activities in developing countries, a figure that may rise to 60 percent by the end of the next century. (These proportions are reversed, of course, for the developed world.) Thus, both issues of economic growth for industrial countries and development in the Third World will be at stake as possible responses to global warming are fashioned.

This examination of the present and future causes of the greenhouse effect reveals the manifold causes and the range of policies that could make some difference in the amount or rate of expected warming. No approach that is narrowly focused on carbon dioxide, for example, or fossil fuels, conservation, or deforestation can fully solve the problem. More important, this look at the vast scope of the greenhouse problem underscores just how deeply its causes are embedded in the central aspects of the world's economic and social activity: its causes cut across transportation, industrial, agricultural, and forestry practices; from the developed to the developing world; and in the very growth of populations and economies. This complexity carries an important implication: although some expected a full solution to the climate change problem to emerge from the negotiations that culminated in the Earth Summit, these talks should be regarded, at best, as a first step in a series of greenhouse negotiations that will likely stretch over decades.

Thus, negotiating and sustaining serious actions to mitigate greenhouse gas emissions will entail real difficulties. While the CFC model has generally been seen as appropriate, the number of significant CFC-producing countries was small. The economic costs, required institutional changes, and affected industries were relatively limited. Those firms that expected to be able to produce CFC substitutes could benefit compared with their competitors and thus could even gain from the treaty. Few of these conditions apply to limits on carbon and other greenhouse emissions. Furthermore, negotiating a broad-scale convention on the apparent causes of global warming will be much more difficult even than the LOS Convention.

A CONVENTION OF LIMITATION VERSUS
A CONVENTION OF EXPANSION

Much of the LOS accord granted or legitimated a series of previously tenuous new claims to ocean resources by many states. Devising an LOS "convention of expansion" involved the relatively easy problem of how to divide an expanding pie.[9] By contrast, climate change negotiations that will have a real, direct effect must focus on working out conventions of limitation, of shared sacrifice, and of painful transfers and compensation—requiring curtailments in energy use, more expensive development paths for developing nations, changes in agricultural patterns, cessation of currently profitable deforestation, and other such activities. To the extent that climate change negotiations are perceived as allocating sacrifices, they will be fundamentally more difficult than the "happier" LOS problem of allocating new resources. Of course, to the extent that the participants focus on the joint gains relative to feared climate disaster, the process will be so much the easier. Some groups that will directly benefit—such as the vendors of renewable, cleaner, more efficient energy and the technologies that make such energy use possible—may join environmental advocates as vocal proponents of a greenhouse control regime.

A TRUE GLOBAL COMMONS WITH DAMAGING INCENTIVES

In 1970 the UN General Assembly unanimously declared deep seabed resources such as manganese nodules to be the "common heritage of mankind;" in fact, this was a statement about property rights to deep seabed resources. By contrast, the global atmosphere is a true commons in the economist's sense that all greenhouse gas emissions from a single country eventually mix and adversely affect the entire world. True commons resources contain economic disincentives for individual initiatives to curb emissions since the full costs of efforts to mitigate harmful emissions by one state must be borne fully by that state, while the benefits of such actions are diffused throughout the global community.[10] Moreover, any benefits of actions now that would slow the present rate of growth of greenhouse gases would only be felt decades later by the inhabitants of a future world. Thus, facing the full costs of abatement today with the prospect of enjoying only a fraction of any future benefits, individual entities have powerful incentives to continue emitting and to "free ride" any costly actions others might take to mitigate the problem. As such, strong political and economic forces can lead states and private parties to postpone any action absent a broad international agreement.

The Bases of Blocking Coalitions in Global Warming Negotiations: Science, Interest, Ideology, and Opportunism

As presently contemplated, the FCCC—which sets forth an agreed definition of the problem and provisions for joint research, monitoring, coordination, and national reporting—is to be followed by specific protocols detailing restrictions to be placed on various sectors. In such an approach, the choice of which specific protocols to pursue singly, in combination, or in sequence (for example, transportation, energy, and tropical forestry) will heavily determine which interests will arise to oppose action; in choosing one's issues, one chooses one's opponents. As elaborated below, this choice of potential opponents—which can be expected to be both private and sovereign, and located both in the developed and the developing worlds—should be a conscious and strategically sophisticated decision. Attention turns naturally enough to opposition that is based on economic self-interest. Yet this is too constricted a view. As the LOS and the CFC experiences attest, scientific disagreement, ideology, and opportunism may also animate blocking coalitions, or nonjoining but opposing entities, that prevent agreement on or implementation of an otherwise desirable treaty.

Blocking coalitions based on economic interest and ideology: The cautionary LOS experience

It is perhaps sobering to recall how the LOS treaty's burdens on seabed mining—for all intents and purposes a nonexistent industry segment—

engendered tenacious and ultimately effective opposition for both prag-
matic and ideological reasons. Major maritime establishments, especially
in the Soviet Union and the United States, were powerfully motivated in
the 1960s by the desire to stop "jurisdictional creep." This phrase refers
to the tendency for territorial claims, especially by coastal and island developing countries, to expand and cast an ever-widening net of restrictions on submarine, ship, and aircraft mobility in what had traditionally
been the high seas. With the likely acceleration of this powerful trend—
to extend territorial sea claims from three miles out to between twelve
and two hundred miles seaward—more than one-third of formerly open
ocean would have the same sovereign status as the land of states. To
navies, this was an intolerable prospect. Thus, the developing world influenced something of high value to the maritime powers.

Emboldened by this genuine maritime interdependence, many developing countries effectively pressed for a seabed regime modeled on the
precepts of the New International Economic Order (NIEO), including
significant wealth redistribution, greater participation by developing
countries in the world economy, and greater Third World control over
global institutions and resources. Real developing country leverage
meant that the maritime powers could not costlessly reject NIEO demands and just walk away. This perceived vulnerability to the power of
the developing countries' coastal states kept the United States and other
maritime powers at the LOS bargaining table for years, but ideological
disagreements ultimately spurred the treaty's rejection.

As the LOS seabed regime took on more of an NIEO-like character, industry opposition grew. The most effective vehicle that industry found to
oppose the treaty was less its economic self-interest than the ideological
cast of the emerging regime. Elements included the declaration that
seabed resources were the common heritage of mankind (seemingly col-
lectivist), seabed production controls (that is, OPEC-like cartels), mandatory technology transfers (negation of intellectual property rights), financial requirements (globally levied taxes), new voting schemes (more like
the UN General Assembly), and the creation of international mining enterprises (worse even than state-owned enterprises). A number of these
issues—such as developing countries' demands for technology and resource transfer and demands for new institutions—are similar to those
now animating climate change negotiations.

Richard Darman, once the vice chairman of the U.S. LOS delegation
and subsequently a senior policy advisor in the Reagan White House
and director of the Office of Management and Budget during the Bush
administration, contended in an influential *Foreign Affairs* article that

"the most important issues at stake in the deep seabed negotiations, however, are not merely questions of manganese nodule mining. What is fundamentally at stake is a set of precedents with respect to systems of governance." In particular, he distinguished between the "precedential elements of the *seabed regime* (as distinguished from *seabed mining*)."[11] The Reagan administration generally concurred. Seabed mining was only a small part of the LOS treaty, but the blocking coalition of seabed miners and policy skeptics that it engendered in the United States was ultimately successful, prevailing over the defense and environmental interests that were the strongest supporters of the LOS Convention.

U.S. negotiating behavior throughout the deliberations of the INC and into the Earth Summit was likewise animated by a sizeable and parallel ideological component: U.S. negotiators were suspicious of multilateral institutions and proposed measures of a nonmarket nature, were hostile to actions that might seem to kowtow to a demanding Third World, and were negative on perceptions of environmental extremism. Much of this ideological animus was ascribed by environmental advocates to the personality and convictions of Bush administration figures such as John Sununu, Richard Darman, and Dan Quayle, but the concern should be focused on more than a few individuals or a single administration.

Undoubtedly, the ideological tenor of the U.S. participation in further climate negotiations will shift in a Clinton administration from its expression during the Reagan-Bush years, but such considerations will very likely continue to play a significant role in these talks for a very basic reason: as with the LOS Convention or the Montreal Protocol, long-term success is impossible without the cooperation of the developing world. Greenhouse gases in the atmosphere are now mainly attributable to past and present activities of developed nations. However, with projected population and economic growth in the developing world, the source of the greenhouse problem will rapidly shift over time, especially if India and China choose their least-cost development paths that rely on their vast coal resources. China, for example, now plans to expand its coal consumption fivefold by the year 2020, a result that would add nearly 50 percent to current worldwide carbon emissions.[12] Anti–global warming steps agreed and taken by the developed world alone could be heavily offset over time by inaction in the developing countries; by the year 2050, it is projected that global warming without developing country cooperation will be 40 percent higher than with it.[13] Thus, the developed world cannot solve the climate problem in the long run without the cooperation of developing nations.

Given the prevailing levels of distrust—not to mention the steep energy requirements vital to development—a threat by key developing nations not to cooperate with an emerging climate regime could have a clear rationale and a measure of credibility, even if such steps are ultimately mutually destructive and even if their effects might be more severe in the developing world. No wonder that, in the words from a recent discussion of climate change and overall Third World concerns, "the problems presented by climate change also present opportunities to reexamine and correct many of the underlying problems of development that have led to the current dilemma . . . including trade issues, debt, technology transfer, technical assistance, and financial assistance."[14] To Southern diplomats who hold this view, the climate change issue may prove to be a very potent bargaining lever, with application well beyond the climate context. According to another observer, "this group sees environment as the same kind of issue in the 1990s that energy was in the 1970s. They hope that the developed countries' interest in the environment can be used over time to wring concessions on development issues from the North."[15]

The fundamental confliction of North-South agendas and the reality of mutual dependence have found and will continue to find expression in climate negotiations.[16] The underlying ideological template, also present in the LOS and Montreal negotiations, is that of the NIEO. A great deal of the preparatory negotiations for the 1992 Earth Summit focused on generalized North-South concerns expressed in well-worn NIEO terms; the Earth Summit itself was barely able to find common principles between a highly negative United States and the North-South cold warriors from the developing countries.[17] The risk is that attempted use of real Southern leverage on behalf of NIEO precepts might meet enduring Northern intransigence that is based on antipathy to the underlying ideology. Parties on either or both sides of this divide could block sustained effective action.

BLOCKING COALITIONS BASED ON SCIENCE AND INTEREST: THE CFC NEGOTIATIONS

Although the CFC accords indeed represent important international coordinating steps, they illustrate bases of potential blocking coalitions, including both scientific disagreement and economic interest, complementary to those explored above in the LOS context. Despite periodic intense public concerns over fears that the supersonic transport and aerosols would deplete the ozone layer, the actions of a relatively small number of industry players (Dupont and Allied in the United States and Imperial Chemical Industries and others in Europe, along with policy skeptics in the major countries) were able to delay action on an ozone convention for a number of years. To understand why, it is critical to focus on internal (domestic) considerations along with considerations in the external (international) negotiating forum.

It is both instructive and sobering to see how CFC industry opposition was overcome by 1987. In part, it was a matter of science. Though predictions of individual scientists varied greatly, consensus estimates of the extent, likelihood, and danger of ozone depletion had declined from the early 1980s prior to the surprise discovery of the Antarctic ozone hole in 1985; thus, industry opposition to regulation during this period had a scientific basis. However, Dupont was publicly committed by statements of company officials to the U.S. Congress to the effect that, if scientific evidence conclusively showed adverse health effects, it would no longer produce CFCs; this declaration was a key factor in Dupont's "conversion."

However, two other dynamics may have been at work in overcoming Dupont's effective blocking actions. First, though it put the work on hold for a time in the early 1980s, Dupont had been intensively engaged in the search for CFC substitutes and appeared to be well ahead of its global competitors in this regard. If this were so, international limits on the amount of CFCs that could be produced and consumed would both permit the price of the allowed production to be raised and place Dupont in a favorable competitive position. Second, as public awareness culminated in tremendous concern about the Antarctic ozone hole, prospects grew substantially for U.S. legislation that would have unilaterally restricted CFC production and use. From Dupont's point of view, while no regulation would have been the preferred alternative, international rules that constrained the entire global industry were far preferable to a U.S. law that singled out domestic companies. Thus, the unusual confluence of several distinct factors—scientific evidence coupled with prior public statements by the company, competitive dynamics within the industry driven by CFC substitutes, and the threat of domestic legislation—were sufficient to turn Dupont around and open a split in the ranks of global industry.

EXTENT OF LIKELY BLOCKING COALITIONS IN ANTI–GREENHOUSE GAS NEGOTIATIONS

These LOS and CFC accounts illustrate how potent greenhouse treaty opponents may be on scientific, economic, and ideological bases. After all, the LOS treaty was scuppered in the United States and in other important industrial nations by the economic and ideological concerns of an industry segment (seabed mining) that did not even exist. With respect to the ozone experience, the 1990 *Economic Report of the President* estimated the U.S. costs of compliance with the Montreal accord at $2.7 billion—one measure, since reduced, of the costs motivating skeptical policymakers and corporate opponents of the treaty.[18] Despite public concern over the ozone layer, the Montreal treaty was effectively delayed for several years by these groups until the scientific consensus shifted.

Although $2.7 billion is certainly a high cost, the same report cited the U.S. costs of a 20-percent cut in carbon dioxide at between $800 billion and $3.6 trillion.[19] If these figures are even remotely accurate, they suggest that those concerned about the prospect of large-scale greenhouse control (for example, policy skeptics, coal and oil companies, and auto makers) would have an economic motivation for opposition, regardless of the level of environmental benefits, literally hundreds of times stronger than that of the CFC industry. The battle throughout the 1980s over amendments to the Clean Air Act, with annual costs in the comparatively mere $25–35 billion range, gives another sobering point of comparison. Cost estimates of the magnitudes mentioned above are by no means universally accepted; respectable analyses suggest that some reductions may be achieved at low or even negative cost.[20] However, it is the credible prospect of burdensome costs that will engender opposition, especially among risk-averse firms that fear they will bear the costs. Furthermore, since the benefits are uncertain, diffuse, and will mainly accrue to future generations, today's opponents are likely to speak with the clearest voices.

Indeed, the powerful coalitions that will arise to resist major greenhouse action are currently fairly quiet (although coal and oil producers have certainly made their views known). They will certainly awaken to the extent that the prospect of such action becomes more likely and that the feared costs are large. Look, for example, to Canada, a country in the rhetorical vanguard of greenhouse concern. If serious actions are proposed, however, will the Canada that pumps oil, cuts forests, and builds cars really just go along? And are those Brazilians who profit from burning rain forests today readily going to buy arguments about future world benefits? More broadly, blocking coalitions are just as likely to arise in Southern countries, whose development could be impeded by anti–greenhouse gas measures, as in the developed countries, whose industries and consumers could face heavy costs. Oil producing states have, of course, lined up in powerful coalitions against a climate deal. Likewise, the imperative for Eastern Europe to grow to consolidate its political gains will weigh against major greenhouse action. Such coalitions will likely be composed not only of traditional nation-states but also of domestic interest groups and transnational alliances. In short, the potentially huge costs that are feared to result from significant anti–greenhouse gas policies offer one measure of the economic motivation for opposition to action and a partial guide to the strength of likely blocking coalitions.

This implication has particular force with respect to the negotiation of national targets, or reductions from given emissions levels that would collectively be within an overall world reduction target. Emissions targets and timetables have been the dominant theme in international discussions over a greenhouse control regime; environmental advocates and media observers have generally judged the seriousness of national governments by their willingness to endorse binding targets and timetables and therefore judged the Brazil framework convention a failure because of its absence of binding commitments. In particular, given the high level of public concern about the greenhouse issue, many environmental advocates expected quick negotiations and decisive agreement on targets. The significant number of industrial countries that unilaterally or in small groups had committed by early 1992 to greenhouse gas stabilization or reduction targets was in line with this optimistic view (although there is a long road between target and result). Yet U.S. (and OPEC) opposition to an overall target (that is, limiting greenhouse emissions in the year 2000 to 1990 levels) effectively kept targets out of the climate change agreement that was signed at the Earth Summit, although the United States later agreed to a greenhouse target.

U.S. opposition to targets may appear anomalous, especially given that all other OECD countries except Turkey had agreed to stabilization by mid-1992 (and the United States had concurred by mid-1993). *Yet as the effects of targets become more specific and stringent, more resistance will grow from those affected.* This correlation implies that negotiating meaningful anti-greenhouse action is likely to take considerable time. The above analysis spells out the extent to which climate change negotiations could seriously impinge on a range of vital activities—far more than the twelve-year LOS process. The much simpler CFC negotiation process, from which specific country obligations emerged, took over five years from the start of negotiations and over ten years from the announcement of the UNEP's 1977 Action Plan to Protect the Ozone Layer. Similarly, the twelve-nation EC Large Combustion Plant Directive to limit acid rain took five years of negotiations, often twice-weekly and among a relatively homogeneous group, to agree on targets.

More recently, and ominously, although the EC as a whole agreed to stabilize its overall greenhouse gas emissions at 1990 levels by the year 2000, its internal negotiations over which nations would be required to make what reductions utterly broke down. This failure should be especially sobering to proponents of targets given the EC's high level of greenhouse concern and its relative homogeneity (especially compared with the broader UN membership that is charged with negotiating the protocol phase of a global climate treaty). With this failure to negotiate country-specific targets, EC attention then shifted to imposing a carbon-related tax. As this alternative was being developed, the *Economist* observed that "the proposed carbon tax has been subject to some of the most ferocious lobbying ever seen in Brussels."[21] Carlo Ripa di Meana, then the EC environment commissioner, charged that the EC faced "a violent assault from industrial lobbies and the [oil-producing] Gulf countries, which even threatened to break off diplomatic relations" following the announcement of the energy tax.[22] Largely as a result of industry opposition, both energy-intensive industries and major exporters were preemptively exempted from the tax before the carbon tax was even proposed as a directive to the Council of Ministers. Furthermore, rather than apply the tax unconditionally as a means of reducing EC carbon emissions as environmental advocates had urged, the tax was made conditional on comparable action by the EC's main trading partners. Similarly, before the Earth Summit and amid some fanfare, Japan accepted targets to cut its carbon dioxide emissions to 1990 levels by the year 2000 and to eliminate ozone-depleting substances. Yet by June 1993, a different picture was emerging. By that time, legislation designed to discourage industries from generating large amounts of greenhouse gases had been watered down in response to strong opposition from business leaders and the Ministry of International Trade and Industry. In particular, legislation would not require environmental impact assessments, penalties, or taxes to discourage polluters.[23] These episodes stand as testament to the power of potential blockers in the realm of climate negotiations and should be far more worrisome than the image of a powerful individual single-handedly preventing climate action.

At first, the acceptance of stabilization targets by all of the OECD countries except for Turkey and the United States might seem to contradict the above analysis, which points out the extent and power of potential blocking coalitions However, another interpretation is possible: as illustrated by the EC experience, targets may be relatively easy to adopt but difficult to implement. One might even draw the analogy to the Gramm-Rudman antideficit law, which eerily resembles a climate framework convention in that it contained targets and timetables but left specific agreement on cuts and tax increases for later. As such, this law served for years at a time of intense public concern about the deficit, as an expedient political solution that allowed executive and legislative officials to declare the deficit problem solved and to return to budgetary chicanery. Similarly, following the second oil shock in 1979, member nations of the International Energy Association agreed on targets and timetables for dramatically reducing their oil imports from OPEC; the results are hard to discern today. It is quite possible that the significant number of unilaterally adopted greenhouse gas control targets or a very weak framework convention that was politically touted as the solution to global warming could have analogous effects. In short, the more clearly identified the objects of anti–greenhouse gas measures—such as targets or carbon-related taxes—the greater the likely opposition from those parties and the more likely that, if adopted, the measures will not be implemented without a far broader and deeper scientific and public consensus on the problem.

In summary, *although economic reasons are most often cited as the basis for opposition to greenhouse action, this is too narrow a view; scientific disagreement, ideological clash, and opportunistic use of apparent bargaining leverage are also likely to play roles.* In principle, each type of blocking coalition might be dealt with according to its basis; in practice, the bases are likely to be intertwined. (The elements cited above are not the only bases for opposition; for example, conflicting values or different attitudes toward risk or the passage of time may engender opposition.) The seabed mining industry appealed to economic interest and ideology in opposing the LOS treaty; science and self-interest played complementary roles in delaying a CFC accord; ideological clash and opportunism may well combine in further global climate talks. Opposition that is really based on one set of reasons will often masquerade behind another, perhaps more politically palatable, one.

OVERCOMING BLOCKING COALITIONS THROUGH CLIMATE NEGOTIATIONS

In light of this exploration of potential blockers, I return to the basic choice posed at the beginning of this chapter. One option would be for negotiators to focus their energies directly on achieving binding restrictions on greenhouse-unfriendly policies—a frontal attack, so to speak, on the forces that thwarted universal adoption of targets and timetables in the last session of the INC. In particular, the goal would be to tighten the framework convention with new negotiations over specific restrictive protocols. Another option would emphasize a more indirect route, using the negotiating process itself and the agreements reached as instruments for nurturing a far more broad-based and deeper consensus among scientists, policymakers, and the public on the need for gaining control of greenhouse gas emissions.

Of course, these options are not mutually exclusive, but the latter approach is more suited to a view that considers the scale of the climate change problem as unprecedented and that regards the role of the negotiations themselves and any agreements they produce as important but only partial responses to the problem. In this interpretation, any solution to the problem of global warming will involve the accumulation of many disparate influences operating at international, national, regional, and local levels. The task of negotiators will be to ensure that the negotiating process and any agreements reached will enhance public discussion and education about climate issues, mobilize decentralized responses, especially among NGOs, as well as support and stimulate worldwide scientific involvement in the search for a deeper understanding of the phenomenon and the nature of effective responses.

RECOMMENDATION: DESIGN THE NEGOTIATING PROCESS TO ENHANCE AND TAP THE VAST POTENTIAL OF ACTIONS SHORT OF INTERNATIONALLY AGREED EMISSIONS LIMITS OR SPECIFIC GREENHOUSE GAS CONTROL REGIMES

Instead of immediately seeking a traditional control regime, other approaches can partly sidestep and prevent the problems of blocking coalitions as well as some of the time lags and sovereignty difficulties characteristic of formal protocol negotiation, ratification, and implementation. For example, former UNEP Deputy Executive Director Peter Thacher has argued against the conventional wisdom of waiting for a negotiated framework convention as a first step that is to be followed by specific negotiated protocols. Instead, in line with experience of the Mediterranean and Ozone Action Plans, he suggested that as many countries as are now willing should first agree on a greenhouse action plan that contains no formal obligations, but that offers the willing sponsors a vehicle within which to promptly commence valuable research, monitoring, and assessment programs as well as to offer developing countries needed assistance to participate in technical and negotiating forums.[24] Such voluntary actions would support and may well speed up the protocol negotiations.

A slightly "harder" option has been suggested by Abram Chayes in an analogy to the launching of the IMF.[25] By creating a post–World War II transition period during which treaty members could simply maintain various forbidden restrictions until they voluntarily relinquished them, the necessary institutional apparatus was developed, professional staffs and reporting practices established, and momentum built toward a result that

was ultimately widely accepted. Applied to the greenhouse case, this scenario would permit further collection of detailed statistical series on global
emissions, facilitate technical assistance to environmental agencies (especially in the developing world), permit the development and empirical val-
idation of more specific performance criteria, and help develop a technically competent and credible monitoring and compliance capability.

Peter Haas and Emmanuel Adler recently edited and contributed to an
issue of *International Organization* (winter 1992) that contained numerous
case studies of the formation of so-called epistemic communities, or
transnational coalitions of experts reasonably close to the decision
processes in various countries who share common views of a problem and
who come to exercise great influence on the international response. One
view is that a very useful function for subsequent climate negotiations
would be to foster the development of an increasingly self-conscious epistemic community around issues of climate science. At present, there is
consensus on much of the basic science of the greenhouse problem, but
the critical issues of timing, magnitude, and distribution still are unresolved. The two key ingredients that might be provided by upcoming negotiations are resources, especially to enable the sustained participation of
scientists from developing countries, and a public spotlight.

Absent a natural climate catastrophe, it is unlikely that future climate
negotiations will have anything like the public salience of the Earth
Summit, which brought together more than 150 nations, 1,400 NGOs, and
8,000 journalists. Nevertheless, future talks should explicitly seek to
build on this widespread public exposure. Given the potential of global
communications technologies and the efforts of concerned governments
and interested NGOs, future climate negotiations themselves and the
public awareness they stimulate can help to spur informal control
regimes, in part by building on and influencing domestic opinion, which
is often led by the actions of NGOs. (The most striking example of this
phenomenon probably occurs in the area of human rights.) The national
reporting requirements contained in the current climate convention—if
beefed up and properly funded—provide a natural vehicle for involving
and mobilizing citizens and advocates. In turn, stronger informal regimes
may come to be embodied in more potent formal instruments that might
earlier have been blocked by opposing coalitions.

Arguably, enough countries and environmental organizations are already sufficiently supportive of greenhouse gas control actions that they
should not have to wait for the conclusion of protocol negotiations to take
meaningful action. In effect, the climate convention signed at the 1992
Earth Summit, which contained no binding greenhouse gas reduction requirements for signatory nations, adopted a version of the approach

sketched above and postponed negotiations over actual restrictions to a later protocol stage. A negotiating process and outcome that did not build on the framework now in place would be publicly invisible and exclusive. A negotiating process and outcome that did build on the framework would be designed for public visibility and inclusiveness as well as for an extension of the scientific consensus on the problem both to new areas of the science and to a broader range of scientists worldwide. In short, if successful, the negotiating process would aim to channel resources and energy into activities that would broaden and deepen the scientific and political coalition in favor of substantial anti–greenhouse gas action.

A second group of recommendations returns to the conventional protocol negotiating process and offers a number of ways of enhancing the prospects for success.

RECOMMENDATION: CHOOSE THE SUBJECT AND NATURE OF LATER PROTOCOLS WITH GREAT CARE

The choice of protocols and the negotiating relationship that is envisioned among them is of central importance; after all, with the choice of a protocol comes a set of opponents (as well as supporters). Protocols have been suggested, in some cases without much explicit analysis of their implications for negotiating success, on a virtually endless number of potential subjects: targets for reducing national greenhouse gas or carbon emissions, credits for providing carbon sinks, automotive transportation, industrial energy use, tropical forestry, agricultural practices, sea level rise, technology transfer, international funds to aid developing countries, a carbon tax, tradeable emissions permits, methane, and so forth.

While it is beyond the scope of this chapter to develop and justify a specific agenda for this process, the choice of protocols should maximize the substantive desirability and the potential of the chosen issue to contribute joint gains to a broad-based group of adherent countries while reducing the likely opposing interests that will be stimulated. Following substantive value, a prime consideration in the choice of protocols should be a clear-eyed view of the likely opposition. Is a proposed target concentrated or diffuse? Is it politically influential in key countries or not? Are the necessary changes inexpensive or very costly?

RECOMMENDATION: MINIMIZE THE RISK OF ENERGIZING AND UNIFYING DISPARATE INTERESTS INTO A LARGE BLOCKING COALITION

A good way to guarantee an endless negotiating impasse would be to handle all of the above-mentioned protocols in a Law of the Atmosphere package to be agreed by consensus. Comprehensive anti–greenhouse gas

efforts that affect a number of potentially powerful interests risk ener-
gizing and unifying otherwise independent, blocking forces. A protocol
that, for example, explicitly targeted oil companies, coal-mining interests,
or automobile manufacturing firms, as well as various agricultural concerns—let alone the full range of human activities that result in greenhouse gases—would almost certainly take a very long time to negotiate
and might never surmount the solid wall of opposition it could raise.[26]

In the greenhouse case, a wise course of action may be to proceed se-
quentially with protocols to avoid the creation of a potent unified opposing coalition. Not entirely tongue in cheek, it may be best to pick
"easy" subjects first to generate momentum—protocols directed at greenhouse contributors that are politically weak, morally suspect, and concentrated in highly "green" countries—with later protocols strategically chosen to build on early successes.

A widely discussed greenhouse control regime involves the allocation
of a number of tradeable emissions permits, such that the overall level of
greenhouse gas emissions could be limited. Beyond the initial allocation,
the ultimate distribution of the permits would not have to be negotiated
or bureaucratically determined since these permits could be bought and
sold. In theory at least, the permits would end up in the hands of those entities that could reduce emissions most efficiently. An ongoing question
with respect to a tradeable permits regime is whether it should only cover
carbon dioxide emissions or whether it should extend to other greenhouse
gases, such as methane and nitrous oxides, in order that the overall least-
cost control actions be chosen. A full answer to this question depends on
issues such as source identifiability, monitorability, and negotiating complexity. From the standpoint of blocking coalitions, however, it is clear
that seeking to negotiate a more comprehensive regime would also risk
unifying a much wider set of disparate, opposing interests. Analogous rea-
soning applies to other proposed anti–greenhouse gas regimes such as
outright emissions limits and various forms of carbon taxes.

RECOMMENDATION: EXPLOIT THE POTENTIAL
OF INCREMENTAL APPROACHES

Beyond measures to prevent the formation of blocking coalitions in the
first place, a number of other approaches can be characterized as incremental. The idea behind them is to gain agreement on a relatively weak
or nonspecific treaty or plan of action in the expectation that, over time, it
will progressively be strengthened. This approach may be a conscious initial choice or it may simply reflect the strength of opposing forces in the
early negotiations. Advocates may settle for what they can get in the hope

that they have set the stage for another round that will conclude more in line with their preferences. This section considers two incremental approaches in rough order of the specificity and weight of the commitments that would be undertaken.

The first incremental approach involves a "baseline protocol." In the best of circumstances, a great deal of valuable time may be lost as countries wait until the international process concludes before taking actions to mitigate greenhouse problems. Some domestic opponents of action in different countries will cynically argue for delaying domestic action until all countries have agreed on reductions. Others will merely regard delay as a prudent bargaining technique to hold off any unilateral action until an international accord is reached. Either way, their blocking (and delaying) potential can be damaging.

One approach to this problem would be the early negotiation of a protocol specifying an early baseline date—perhaps a date in the past—after which anti–greenhouse gas measures taken by individual countries would be credited against the requirements of a later international agreement.[27] With such a date agreed, states could promptly undertake unilateral or small-group initiatives to reduce greenhouse gas emissions in the confidence that these measures would count toward the reductions required by an ultimate regime. Such a baseline-year agreement, perhaps negotiated as a protocol, could help to neutralize a major argument of domestic opponents of anti–greenhouse gas measures who hold that action without an overall international agreement is either unwarranted or foolish.

Given the likely time required for an overall agreement embracing substantive anti–greenhouse gas measures such as binding targets and timetables, a preliminary baseline protocol of this sort should prove far easier to negotiate quickly. Incidentally, such a baseline protocol need only assure states that their actions subsequent to the agreed baseline year would count; the question of the status of actions taken prior to the agreed date could be explicitly left for future negotiation. (A rough U.S. analogue to this approach is contained in the 1992 energy strategy bill that permits companies' anti–greenhouse gas measures to count against future regulatory requirements.)

The second incremental approach involves ratchet mechanisms. Suppose that greenhouse gas reduction targets were set at extremely modest levels in an initial protocol. Likewise, imagine that an international tax on carbon emissions were initially set at a very low rate—for example, to collect resources for an international environmental fund. Given its low rate, this tax (or set of reduction targets) might not trigger concentrated opposition. Later, with the monitoring and collection structures in place, the tax rate (or targets) might be ratcheted up if the state of the science merited it and if broad-based support for such a move existed.

Indeed, a review of the history of the ozone negotiations suggests the potential value of such a ratcheting device. When an agreement to set CFC limits proved unreachable in 1985, the United States and others pressed for the Vienna Convention that collectively legitimated the problem, set in motion joint efforts at monitoring, coordination, and data exchange, and envisioned the later negotiation of more specific protocols.[28] In 1987, after scientific consensus on the problem had solidified and industry opposition was largely neutralized, the Montreal Protocol embodied an agreement to cut CFC production and use 50 percent by the year 2000. Many environmental activists harshly criticized these agreed targets as inadequate.

However, the institutional arrangements set up by the Montreal Protocol included provisions to facilitate a review of the agreed limits in the face of new evidence (or, effectively, with shifts in public opinion). In effect, these provisions functioned as a ratchet, whereby later findings such as the direct link between CFCs and the ozone hole would stimulate treaty parties to tighten the limits over the 50-percent base. As the UNEP's Mustafa Tolba recently put it, "By aiming in 1987 for what we could get the nations to sign . . . we acquired a flexible instrument for action. If we had reached too far at Montreal, we would almost certainly have come away empty-handed. . . . [The] protocol that seemed modest to some . . . is proving to be quite a radical instrument."[29] This assessment was borne out by the 1990 London negotiations that converted a 50-percent reduction into a virtual CFC ban. This model of settling for relatively modest restrictions on which early agreement can be reached, together with arrangements that facilitate reconsideration, may well be emulated in the climate context.

There is, however, a danger to partial agreements, as exemplified by the 1963 Partial Test Ban Treaty. A number of observers have criticized these accords as stopping too soon and wasting the intense public pressure for change, when, arguably, a comprehensive test ban treaty was then attainable. By negotiators' addressing concerns about strontium 90 from atmospheric testing in the food chain (in mothers' milk in particular), this argument goes, the broader dangers of nuclear testing were not addressed and a valuable opportunity was squandered. Rather than acting as a stepping stone to a larger accord, the Partial Test Ban Treaty became a stopping place. (Recall also the analogy to the U.S. Gramm-Rudman anti-deficit law that was drawn above.)

With respect to climate change negotiations in particular, it is quite likely that public concern will be cyclic, in part as a result of natural climate variability as well as unrelated environmental events (such as medical waste on beaches and the Exxon *Valdez* oil spill). Arguably, a naturally occurring period of climate calm, including milder summers and normal rainfall, will lead to reduced public concern and pressure for action. Moreover, scientific understanding will change over time. These prospects argue for more limited agreements, with analogues to the ratchet mechanism in the Montreal Protocol, if and when more stringent action appears warranted. Such agreements could constitute a "rolling process of intermediate or self-adjusting agreements that respond quickly to growing scientific understanding."[30] In addition, an even more fundamentally adaptive institution might be envisioned that better matched the rapidly changing science and politics of this set of issues.

RECOMMENDATION: BE CAUTIOUSLY OPEN TO LINKAGES AMONG ISSUES (AND/OR SPECIFIC PROTOCOLS)

In the face of substantial challenges, a successful accord on climate change calls for a process designed to achieve results that can be sustained over time and modified as appropriate. In particular, a process like the ones that took place in Vienna and Montreal, with independent protocols to be negotiated on a step-by-step basis, was thought to have the advantage of speed and relative simplicity over a comprehensive LOS-like approach. This raises the more general question of how to deal with greenhouse issues (or protocols): singly, comprehensively, or in intermediate-sized linked packages. The answer, most usefully explored in the LOS context, has a direct implication for ensuring enough gains in an agreement to attract a winning coalition.

Many factors contributed to the lengthy LOS process, but four procedural cornerstones virtually guaranteed its duration and, if these procedures are adopted, could easily do the same to global warming negotiations. These factors include (1) almost universal participation, combined with (2) a powerful set of rules and understandings aimed at making all decisions by consensus if at all possible, (3) a comprehensive agenda, plus (4) the agreement to seek a single convention that would constitute a package deal.[31] The rationale for each of these components was understandable; however, in a climate context *a universally inclusive process with respect both to issues and participants, together with the requirements of consensus on an overall package deal, would be very time-consuming, thus holding the ultimate results hostage to the most reluctant party on the most difficult issue.* In practice, the LOS conference was less constrained by absolute adherence to these procedural choices, but the powerful bias that made the process move at a snail's pace was very real.

Reacting against the LOS approach (that is, a comprehensive agenda with the requirement of a package deal), climate change negotiators aimed for a framework convention to be followed by specific protocols. In line with the CFC experience, this approach retained the aims of universality and consensus, but dropped comprehensiveness and the goal of a package deal in favor of single, separable protocols on limited subjects. This alternative has attractive negotiating features, but it is worth noting that it was the failure of precisely this approach—negotiation of separate miniconventions, which are analogous to protocols—in earlier LOS conferences (in 1958 and 1960) that indirectly led back to the comprehensive package approach of the 1973 LOS conference.

THE PROBLEM OF SELECTIVE ADHERENCE

The LOS experiences in 1958 and 1960 suggest that sometimes issues must be linked. By 1958, for the First UN LOS Conference, the International Law Commission had suggested a negotiating structure with four separate conventions concerning different issues, such as the breadth of the territorial sea and the extent of the continental margin. With respect to the comprehensive agenda of the 1973 LOS talks, Conference President Tommy Koh observed that

> A disadvantage of adopting several conventions is that states will choose to adhere only to those which seem advantageous and not to others, leaving the door open to disagreement and confrontations. The rationale for this [comprehensive] approach was to avoid the situation that resulted from the 1958 conference which concluded four [separate] conventions.[32]

Such an uneven pattern might also result from a framework-protocol structure on climate change. Imagine Libya signing a forestry convention while Nepal agreed to a transportation and automotive protocol. For individual countries or groups of similar ones, a single issue often represents either a clear gain or a clear loss. As with the early LOS conferences (with independent miniconventions), countries sign the gainers and shun the losers. In a climate context, for example, China may resist a specific fossil fuel protocol that would place restrictions on the development of its extensive coal resources.

LINKAGE FOR JOINT GAINS AND BREAKING SINGLE-ISSUE IMPASSES

Such single-issue protocols may prove nonnegotiable unless they can be combined with agreements on other issues that offset the losses, or that at least seem to distribute them fairly. A package deal may offer the possibility of trading across issues for joint gain, thus breaking impasses resulting from treating issues separately.

For example, following the 1958 and 1960 LOS experiences, two separate negotiations were attempted; until linked, each proved fruitless. With deep seabed resources considered the common heritage of mankind, the Seabeds Committee undertook a negotiation over the regime for seabed mining. The developing countries wanted this convention to offer meaningful participation in deep seabed mining and the sharing of its benefits. However, the developed countries—whose companies potentially had the technology, the capital, and the managerial capacity to mine the seabed—saw no reason to be forthcoming, and these negotiations were inconclusive. At about the same time, strenuous efforts by the United States, the Soviet Union, and other maritime powers, who were greatly concerned about increasing numbers of claims by coastal, strait, island, and archipelagic states to territory in the oceans, sought to organize a set of negotiations that would halt such creeping jurisdiction. In effect, the maritime powers were asking coastal states, without compensation, to cease a valuable activity (that is, claiming additional ocean territory). Not surprisingly, these discussions over limits on seaward territorial expansion in the ocean yielded scant results.

Seen as separate protocols, these two issues, taken independently, were not susceptible to agreement. Ultimately, however, the linkage in bargaining of these two issues, navigation and nodules—together with concerns over the living resources and outer continental shelf hydrocarbons—came to be at the heart of the comprehensive LOS negotiations. With respect to climate change negotiations, it is easy to imagine that separate protocols calling on different groups to undertake painful and costly measures will similarly be rejected unless they can be packaged in ways that offer sufficient joint gains to key players. Since any action on climate change will largely involve shared and parallel sacrifice, it is probably only by a linkage of issues such as technological assistance and various forms of financial or in-kind compensation that many developing countries will be induced into joining. As such, one should expect great pressure toward combining issues that might initially be conceived as separate protocols for purposes of negotiation.

Certain classes of issues, of course, have been inextricably linked in the negotiations so far. In particular, the question of how greenhouse gas control measures among the other items in the Earth Summit's agenda are to be financed was the subject of furious negotiation in the INC and at the Earth Summit. While no specific commitments were made by the North, the prospect of substantial resources being funneled through the World Bank proved to be critical in moving the negotiations forward. Almost independent of how the Global Environmental Facility negotiations proceed, donor country support for an additional "earth increment" to World Bank financing resources (the "tenth replenishment" of the International Development Association) will be seen as critical.[33] Such linkages, perhaps external to the actual climate negotiations, need special attention.

Just as in the LOS experience, mutually beneficial manageable packages of protocols under a framework climate convention might be cautiously extended to other environmental issues that arose in the context of the 1992 Earth Summit. This action might have the effect of bringing on board potential blocking coalitions from the developing world. For example, desertification and soil erosion issues may be more pressing to key developing countries than questions of greenhouse emissions. Many developed countries that are unwilling to offer "bribes" to induce the participation of developing countries may nonetheless be genuinely concerned about and more willing to be forthcoming on these regional issues in the context of a larger agreement that promised global climate benefits. Similarly, more expansive versions of so-called debt-for-nature swaps may be explored. One of the most potent long-term steps that could be taken by developing countries to combat global warming (as well as a host of other environmental issues) would be significantly stepped-up population control programs.[34] Unlike, say, energy-use restrictions, this course of action has the virtue of helping rather than hindering economic development objectives. For cash-strapped developing countries, relatively modest aid from developed countries in this area could considerably enhance domestic population control efforts.

Given this analysis, a central problem in the design of future greenhouse negotiations would seem to be finding a constructive path between the Scylla of a comprehensive package agenda that risks LOS-like complexity and the Charybdis of independent, single-issue protocols that may lack sufficient joint gain and thus risk selective adherence.[35] Rather than trying to predict the appropriate linkages, the conference should be designed in such a way as to facilitate linkages as they become evident and necessary. It is generally preferable for negotiators to deal with issues on their separate substantive merits as much as possible yet be alert to potential linkages in order to break impasses. This scenario suggests a conference design with independent working or negotiating groups with a higher level body that seeks to integrate issues across groups and facilitate valuable, but limited, trades.

However, issues should be linked with caution. It can be extraordinarily difficult to unpackage them once they have been combined for bargaining purposes. For example, the United States was generally in favor of the navigational portions of the LOS treaty but had problems with the concessions demanded on a seabed regime. It exerted strenuous efforts at unlinking or separating these topics into manageable packages, but to no avail. The package deal was too strong in the minds of many delegates, and ultimately, the LOS Convention contained both elements.

RECOMMENDATION: TAKE ACTIVE STEPS TO AVOID TRIGGERING A HOPELESS NORTH-SOUTH IMPASSE

As discussed above, there is an acute risk that a larger North-South agenda—some of it only loosely related to climate change and much of it highly contentious—will occupy center stage in greenhouse negotiations over time. Indeed, these talks have already been characterized by aggressive demands by developing countries for technology transfer and large resource commitments from the industrial world. It is clear that finance and technology transfers, for example, constitute legitimate interests, but the extent to which developed countries will be forthcoming on them in the context of future climate change negotiation is far less clear—especially given ideological reservations about what could be seen as resurgent demands for a discredited NIEO. Moreover, despite the keen concern in many nations about climate change, the greenhouse problem is speculative, contested, far in the future, and very costly to address merely on its own terms, that is, absent additional resources to mitigate generalized problems of developing countries. The uncertain prospect of global warming may not be a strong enough hook on which to hang a larger North-South agenda. The spectre of a North speaking "environmentalese" and a South speaking "developmentese," with each side talking past the other, is all too real.

With the crumbling of socialist ideology in Eastern Europe and the Soviet Union, many Europeans are also becoming less receptive to formerly attractive NIEO precepts. Thus, if the language negotiated as part of a climate change convention invokes images such as central command, heavy-handed international bureaucracy, forcible technology transfer, blame-casting ideological declarations, guilt-based wealth transfers, and the like, the results of any such negotiation run substantial risk of being overturned. Indeed Northern opponents of a climate change convention, especially those in the United States, may well base their negative stand on the actual or supposed adverse ideological cast of the regime.

Like LOS, therefore, real mutual interdependence means that climate change talks have the ingredients for an inescapable, long-term, North-South engagement: Southern insistence on NIEO-like measures that would meet with Northern resistance. Given that Southern commitment to the NIEO per se has moderated considerably since the 1970s, the risk of an ideologically driven impasse is probably manageable with some conscious effort. Both the INC and the Earth Summit itself offer grounds for optimism, although the issue may only be in temporary remission. As will be discussed below, creative steps are essential to meet the legitimate interests of developing countries while reducing the risks that such an engagement would result in endless delay and damaging ideological confrontation, with no action to address the greenhouse problem or development imperatives.

RECOMMENDATION: AN INFORMATION NEGOTIATING PROCESS TO PARALLEL THE FORMAL ONE.

A number of well-publicized regional workshops in advance of the negotiations—presented by regional scientists and policy figures that focused on possible impacts—could help spread the conviction that global warming is a common threat from a shared problem. Joint research and study between developing and developed countries should likewise be encouraged, perhaps building on the work of the IPCC jointly sponsored by the UNEP and the World Meteorological Organization.

During the negotiations themselves, similar informal educational events could be helpful. One extraordinary element of the LOS experience, which has been detailed by outside observers, consisted of the influence of a computer model of deep ocean mining developed at the Massachusetts Institute for Technology. This model came to be widely accepted in the face of the great uncertainty felt by the delegates about the engineering and economic aspects of deep seabed mining. A critical point in the negotiations occurred during a Saturday morning workshop—held outside UN premises under the auspices of Quaker and Methodist NGOs—in which delegates from developed and developing countries were able to meet and extensively query the MIT team that had built the model. Indeed, over time, the delegates came to make frequent use of the model for learning, mutual education, invention of new options, and even as a political excuse to move from frozen positions.[36]

Similarly, a series of informal, off-the-record workshops, where diplomats and politically active participants in the negotiation gathered, aided the Montreal Protocol process. These events greatly increased mutual understanding, improved relationships, and contributed to a successful treaty. Despite its potential abuse by advocates, outside scientific information—when it is seen to be objective and is accessible to the participants—can help move a complex negotiation, even one that is highly politicized and ideologically controversial, in the direction of mutual cooperation. (Of course, improved science might instead clarify winners and losers, thus polarizing the issue.)

As a broader proposition, negotiations that take place entirely through formal diplomatic means have a diminished prospect of success relative to those that encourage informal interactions, the buildup of trust, and the enhancement of personal relationships. Such a parallel, informal process can be hosted by any number of sympathetic participants and observers, from delegations themselves to NGOs of all sorts. A wise secretariat will seek to provide occasions and venues for such events to flourish.

RECOMMENDATION: AN ENHANCED ROLE FOR ADVISORY GROUPS AND THE FORMATION OF CROSS-CUTTING COALITIONS

Given the actual and feared adverse impacts of the measures under discussion, conference leadership would be wise to continue to make extensive use of broadly constituted advisory groups, composed of business and other multinational interests, in order to understand concerns, anticipate emerging problems, correct misapprehensions, and communicate about the issues and the evolving negotiating responses. Not only could the two-way communication be useful in such settings, but cross-cutting coalitions might form. For example, industries that could gain from substantial anti–greenhouse gas actions in the developing world (for example, by supplying critical technology for energy efficiency) might make common cause with key developing countries and green advocacy organizations in arguing the case for more developed country assistance for this purpose. The potential value of the Commission on Sustainable Development and the Business Council for Sustainable Development is very high indeed.

RECOMMENDATION: ELABORATION AND DIFFUSION OF A NEW IDEOLOGICAL TEMPLATE

The North-South conflict has been a staple of recent global negotiations beyond UNCED—from the UN Conference on Trade and Development to debt and codes of conduct for transnational corporations—although the overt NIEO focus has moderated in the years between the LOS and the Montreal talks.[37] Joint development of a new ideological template within which the climate question could be negotiated offers another means to escape impasse. Such a new conception could avoid lumping countries with vastly different climate interests—from coal-rich developing countries such as China and India to sub-Saharan Africa to the Second World of Central Europe to Norway and the United States—into catchall categories such as North and South. The most promising candidates to date are the principles of sustainable development, which was articulated by the Brundtland Commission in *Our Common Future* and elaborated in the discussion of the Earth Charter and Agenda 21.[38] The principles underlying sustainable development insist on development that meets the needs of the present without compromising the ability of future generations to meet their own needs. Although in need of clearer definition, these widely discussed principles call for tight links between environment and development, for institutions that integrate environmental and economic decision making, for international cooperation on global issues, and for major efforts toward more sustainable paths of population growth and energy and resource use. Whether such principles can come to have the acceptance, weight, and specific implication needed to steer climate negotiations safely away from stale North-South rhetorical exercises remains to be seen, but they provide a promising possibility.

RECOMMENDATION: CONSIDER DIRECTING NEGOTIATING
ENERGIES TOWARD A SMALL-SCALE, EXPANDING AGREEMENT

The complexities of a universal process may still threaten endless delay or impasse. Suppose, however, that a smaller group of industrialized states with potent domestic interests keenly interested in anti–greenhouse gas measures were to negotiate among themselves a reduction regime that included timetables and targets, either voluntary or mandated. Presumably the core group would include major contributors to the greenhouse problem in which there was substantial and urgent domestic sentiment for action. A natural starting core would be the twelve nations of the EC, the six member states of the European Free Trade Association, plus Japan, Australia, and Canada—all of which by 1992 had unilaterally or collectively adopted greenhouse gas stabilization or reduction targets.[39] With the advent of the Clinton-Gore administration, the prospects for more active U.S. participation, support, and leadership certainly improve.

Agreement among such a group would likely prove far easier to achieve than would a global accord because of the fewer number of states involved as well as their greater economic and political homogeneity. The obvious umbrella for such an effort is the Framework Convention signed at the 1992 Earth Summit, but existing institutions (such as the UN Economic Commission for Europe or the OECD) might also facilitate the process. In addition, while there would clearly be substantial negotiating difficulties involved, this smaller scale process could avoid a protracted, inconclusive North-South clash that might stymie a larger forum.

To be effective in the longer term, of course, a smaller scale agreement would have to be expanded later to include key developing countries such as China, India, Brazil, Indonesia, and Mexico, as well as additional developed nations, especially in Eastern Europe. In this sense, an agreement explicitly designed for an increasing number of adherents has strong parallels to agreements that ratchet up targets in order to become increasingly stringent. The design of the smaller negotiation could anticipate and facilitate such an expansion in several ways.

First, the smaller agreement should seek to build on the present framework convention, when the general problem will have been legitimated and accepted to the largest extent possible. Second, it should be cast not as an alternative to the global process over protocols, but as a complement to it in which those nations that have caused the present greenhouse gas problem are those who take early action to mitigate emissions. This process would give the smaller group that had agreed to cuts a higher moral standing when soliciting later reductions from others. Third, the smaller scale group should structure its accord with the explicit expectation of collectively negotiating incentives, likely tailored to special circumstances, for key developing nations to join the accord. For example, the smaller group might agree to tax its members on their carbon emissions. All or part of those tax proceeds could be used to gain the acquiescence of other key countries to anti–greenhouse gas measures. The smaller group could create an entity that would carry out these negotiations with these key countries, rather than leaving such negotiations to ad hoc efforts by individual member countries. The smaller group might also be especially effective in soliciting support for the next replenishment of international development assistance.

Negotiations between the smaller treaty group and, say, China could set a schedule of emissions targets and offer China significant incentives to reach them. Or the group could address a range of China's concerns in return for less climate-damaging development (for example, assistance with greater exploration for Chinese natural gas reserves, Chinese agreement to use CFC substitutes in refrigeration, Chinese agreement to make its coal development more greenhouse-friendly, perhaps by the transfer of more efficient electrical generating equipment). Such customized small-group negotiations with China, India, Brazil, and others should be more conducive to environmentally desirable results than would generalized North-South clashes in a full-scale UN conference.

The fourth option is considerably more ambitious and contentious. As the group of adherents to the smaller convention grew in size, it might choose to impose a tax on products imported into member countries from nonadherents, perhaps based on the direct or indirect carbon content of those products. While this tax would elicit an extreme reaction from GATT, the carrot (providing individually tailored negotiated incentives for nonadherents to join) and the stick (raising such a "carbon fence" around groups resistant to anti–greenhouse gas measures) might together lead to a much larger number of countries jointly taking measures to prevent climate change. Evidently, a price to be faced, deliberated, and accepted by the smaller group would be a substantial number of free-riding countries. With a large enough group of adherents, however, the actions of a smaller group could still be preferable to no agreement at all.

Ironically, although a number of developing countries have joined the Montreal Protocol, it is quite possible to interpret this accord, after the fact, as strongly analogous to the smaller scale convention proposed above. While carried out in the context of a widely accepted framework (the Vienna Convention), the relatively small number of key CFC-producing countries ultimately acceded to the CFC reductions in the Montreal Protocol. However, important developing countries (India, China, and Brazil) did not agree until 1990. India, for example, demanded $2 billion—a number related to its cost of using more ozone-friendly technology in the future—as its price to join the 1987 protocol.[40] In 1990 a number of developed nations agreed to provide such assistance up to $240 million. This amount proved sufficiently attractive to representatives of states such as India and China that they indicated their willingness to join. However, as a result of the smaller scale Montreal Protocol, extremely significant ozone-protection measures are now underway even before the full resolution of important issues concerning financial aid and technology transfer to the developing world has taken place.[41]

CONCLUSION

The problems that negotiators will face in the next phases of a regime to control global warming illustrate the powerful barriers to agreement, versions of which apply in a large number of contexts. This chapter has sought to clarify the nature of these barriers and suggest constructive responses to them. Environmental diplomats have largely taken negative lessons from the LOS negotiations and positive ones from the CFC accords in envisioning a framework-protocol process for global warming. However, gaining significant future action to curb greenhouse gas emissions will be a far more difficult task than either dealing with ocean resources or the ozone layer. Despite the apparent appeal of the step-by-step approach, a review of the evolution of the LOS process from separate miniconventions into a comprehensive treaty illustrates the powerful forces that will likely operate on a climate change negotiation that seeks to combine protocols and to collapse a many-stage process into a more unified effort. The trick will be to find smaller, more manageable packages that embody enough mutual gains to attract key players.

The power of the coalitions that will arise to block greenhouse action—not merely for reasons of economic interest, but also for reasons of science, ideology, or opportunism—must be taken into account in the design of an effective negotiating process. Preventing and overcoming these forces could be aided by a sophisticated selection and sequencing of protocols, as well as by such innovative devices as ratchet mechanisms, negotiated baselines, and voluntary actions short of negotiated targets. Even if these hazards are avoided, the possibility of a North-South impasse looms. A number of actions could mitigate it, however, including workshops, negotiation process choices, creative linkages, and advancement of new ideological templates. If these measures are unsuccessful, attention may shift to a smaller scale, expanding convention that could use incentives and penalties to later bring other states into its fold. Good candidates to start this process include those countries that have unilaterally committed to greenhouse emissions targets.

Advocates need to keep in mind the distinction between measuring success by the number of diplomatic instruments ratified versus actual policy shifts over time. The obvious focus of energy is on the former. Given the sheer scale of the factors contributing to the climate problem, however, negotiated results will at best be one of many factors that accumulate to change widespread and deep-seated behaviors that generate greenhouse gases. Thus, rather than conceptualizing negotiations primarily as potential direct producers of greenhouse restrictions, this chapter has argued that negotiations themselves should also be understood as a potential contributor to broader scale awareness, scientific development, and consensus. The latter effect may be far more powerful than the former. A climate-negotiating process and its outcome should be measured by the extent to which they stimulate public visibility and education, mobilize nongovernmental action, and foster widespread involvement of scientists in the development of a fuller consensus. The opposite of success in these terms would be a process and result that were diplomatically correct but which proceeded invisibly and exclusively. One scarcely looks forward to the international equivalent of the U.S. Gramm-Rudman antideficit targets and timetables.

In summary, to an advocate of a new greenhouse gas control regime, the fundamental negotiating task is to craft and sustain a meaningful winning coalition of countries that will back such a regime over time. Two powerful barriers to this fundamental task are (1) that each member of the coalition fails to see enough gain in the regime, relative to the alternatives, to adhere and (2) that potential and actual blocking coalitions of interests opposed to the regime are neither prevented from forming, acceptably accommodated, or otherwise neutralized. The recommendations for negotiation design developed in this chapter suggest that, over time, as the science and politics warrant, there are many ways to surmount these daunting barriers for climate negotiations, and, one hopes, in other areas.

ACKNOWLEDGMENTS

This chapter draws heavily, with permission, from my chapter in a volume edited by Irving Mintzer (in *Negotiating Climate Change: The Inside Story of the Rio Convention*, Cambridge: Cambridge University Press, 1994, 277–320). This chapter contains an extensive set of background and supporting citations that are incorporated by reference into the present chapter. I am indebted to the same people and organizations acknowledged therein, especially to the Negotiation Roundtable at Harvard and the Salzburg Environmental Initiative, as well as to Kenneth Arrow's subsequent helpful suggestions on a related paper. Support of the Office of Policy and Evaluation of the U.S. EPA, the Charles Stewart Mott Foundation, and the Stockholm Environmental Institute is gratefully acknowledged.

NOTES

1. Negotiation analysis is a prescriptive approach to negotiating situations that draws on game-theoretic concepts but does not presuppose

the full rationality of the participants or common knowledge of the negotiating situation. For expositions, see, for example, Howard Raiffa, *The Art and Science of Negotiation* (Cambridge: Harvard University Press, 1982); David A. Lax and James K. Sebenius, *The Manager as Negotiator* (New York: Free Press, 1986); James K. Sebenius, "Negotiation Analysis: A Characterization and Review," *Management Science* 38, no. 1 (January 1992): 19–38; or H. Peyton Young, ed., *Negotiation Analysis* (Ann Arbor: University of Michigan Press, 1991).

2. In the climate case, blocking coalitions may include nonjoiners and free riders. However, peculiarities in the rules of conference diplomacy may actually allow such nonjoiners to block agreements that are widely desired. For traditional discussions of these coalitional concepts, see R. Duncan Luce and Howard Raiffa, *Games and Decisions* (New York: Wiley, 1957); or William Riker, *The Theory of Political Coalitions* (New Haven: Yale University Press, 1962). Here, winning coalitions are only defined with respect to a set of policy measures from the point of view of a particular actor or actors; such coalitions consist of sufficient numbers of adherents to render the policy effective (again from the point of view of the specific actor or actors). Blocking coalitions are those opposing interests that could prevent a winning coalition from coming into existence or being sustained. The term "actor" should be contextually obvious and can include states, domestic interests, and transnational groupings of either as appropriate. Although the necessary conditions described above are extremely important, sufficient conditions do not in general exist for an agreement to be reached and an impasse or escalation avoided (see Lax and Sebenius, *The Manager as Negotiator*).

3. Climate change was only one of the many subjects taken up at the 1992 conference, which was timed to take place on the twentieth anniversary of the initial UN environmental conference held in Stockholm. The vast agenda of the 1992 conference also included other atmospheric issues (ozone depletion, transboundary air pollution), land resource issues (desertification, deforestation, and drought), biodiversity, biotechnology, the ocean environment, freshwater resources, and hazardous waste. UN General Assembly, "United Nations Conference on Environment and Development" (General Assembly Resolution 228, 44 UN GAOR Supp. [No. 49] at 300, UN Doc. A/44/49, 1989).

4. For a summary of the unilateral and small-group greenhouse gas reduction and stabilization targets adopted worldwide, see *Global*

Environmental Change Report (Arlington, MA: Cutter Information Corporation), 2, no. 19 (9 November 1990): 1–5, as well as subsequent issues.

5. "Clinton Commits U.S. to Greenhouse Target, Biodiversity Convention," *Global Environmental Change Report* (Arlington, MA: Cutter Information Corporation) 5, no. 8 (23 April 1993): 1.

6. Other useful precedents include the Limited Test Ban treaty and nonproliferation agreements, the Basel Convention on hazardous wastes, the Convention on International Trade in Endangered Species, the Antarctic treaty, and various regional environmental accords such as the Mediterranean Action Plan. For useful distillations of some of the lessons from these and many other related accords, see Oran R. Young, "The Politics of International Regime Formation: Managing Natural Resources and the Environment," *International Organization* 43, no. 3 (summer 1989): 349–375; Peter S. Thacher, "Alternative Legal and Institutional Approaches to Global Change," *Colorado Journal of International Environmental Law and Policy* 1, no. 1 (summer 1990): 101–126; and, especially, Peter H. Sand, *Lessons Learned in Global Environmental Governance* (Washington, DC: World Resources Institute, 1990).

7. The following LOS discussion generally relies on Ann L. Hollick, *U.S. Foreign Policy and the Law of the Sea* (Princeton, NJ: Princeton University Press, 1981); James K. Sebenius, *Negotiating the Law of the Sea: Lessons in the Art and Science of Reaching Agreement* (Cambridge: Harvard University Press, 1984); Bernard Oxman, David Caron, and C. Buderi, *Law of the Sea: U.S. Policy Dilemma* (San Francisco: ICS Press, 1983); E. L. Richardson, *The United States and the 1982 UN Convention on the Law of the Sea: A Synopsis of the Status of the Treaty and Its Expanded Role in the World Today* (Washington, DC: Council on Ocean Law, 1989); and E. L. Richardson, "Law of the Sea: A Reassessment of U.S. Interests," *Mediterranean Quarterly: A Journal of Global Issues* 1, no. 2 (spring 1990): 1–13. See current issues of *Ocean Policy News* (Washington, D.C.: Council on Ocean Law) for reports on the total ratifications to the LOS convention.

8. The following account draws generally on Richard E. Benedick, *Ozone Diplomacy* (Cambridge: Harvard University Press, 1991); Benedick, "Ozone Diplomacy," *Issues in Science and Technology* 6, no. 1 (fall 1990): 43–50; Benedick, "The Montreal Ozone Treaty: Implications for Global Warming," *The American University Journal of International Law and Policy* 5, no. 2 (winter 1990): 227–234; David D. Doniger, "Politics of the Ozone Layer," *Issues in Science and Technology* 4, no. 3 (spring 1988): 86–92; and Peter M. Haas, "Ozone Alone, No

CFCs: Ecological Epistemic Communities and the Protection of Stratospheric Ozone" (paper presented at Conference on Knowledge, Interests, and International Policy Coordination, Wellesley College, 1989).

9. There were, of course, limitations on various activities (for example, coastal state seaward territorial claims and marine scientific research) negotiated in the LOS context. Not surprisingly, these were among the most difficult aspects of the conference.

10. G. Garrett Hardin, "The Tragedy of the Commons," *Science* 162, no. 3859 (13 December 1968): 1243–1248.

11. Richard G. Darman, "The Law of the Sea: Rethinking U.S. Interests," *Foreign Affairs* 56, no. 2 (January 1978): 373–395.

12. Michael Grubb, "The Greenhouse Effect: Negotiating Targets," *International Affairs* 66, no. 1 (winter 1990): 75.

13. D. Lashof and D. Tirpak, *Policy Options for Stabilizing Global Climate* (Washington, DC: U.S. EPA, Office of Policy, Planning, and Evaluation, 1989), 40–43.

14. Christopher D. Stone, "The Global Warming Crisis, If There Is One, and the Law," *The American University Journal of International Law and Policy* 5, no. 2 (winter 1990): 497–511.

15. Richard H. Stanley, *Environment and Development: Breaking the Deadlock* (Report of the 21st UN Issues Conference) (Muscatine, IA: Stanley Foundation, 1990), 8.

16. See, for example, Stephen D. Krasner, *Structural Conflict: The Third World against Global Liberalism* (Berkeley and Los Angeles: University of California Press, 1985).

17. The phrase is from Richard N. Gardner, *Negotiating Survival* (New York: Council on Foreign Relations, 1992), 25.

18. U.S. Council of Economic Advisors, *Economic Report of the President* (Washington, DC: U.S. GPO, February 1990).

19. See Ibid., 234, based on A. S. Manne and R. G. Richels, "Global CO_2 Emission Reductions—the Impacts of Rising Energy Costs" (Menlo Park, CA: Electric Power Research Institute, 1990).

20. For a critique of the Manne-Richels estimates, see R. H. Williams, "Low Cost Strategies for Coping with Carbon Dioxide Emission Limits," Center for Energy and Environmental Studies (Princeton University, 1989). More generally, for a sophisticated review of various cost estimates, see William R. Cline, *Global Warming: The Economic Stakes* (Washington, DC: Institute for International Economics, 1992).

21. *The Economist*, 9 May 1992, 19.

22. *Financial Times*, 15 May 1992, 3.

23. "Japan Backing Down from Large-Scale Global Warming Reduction

Plans," *Global Warming Network Online Today* (Alexandria, VA: Environmental Information Networks), 3 June 1993, 1.

24. Thacher, "Alternative Legal and Institutional Approaches."

25. Abram Chayes, "Managing the Transition to a Global Warming Regime or What to Do Until the Treaty Comes," in *Greenhouse Warming: Negotiating a Global Regime*, ed. Jessica Mathews (Washington, DC: World Resources Institute, 1991), 61–68.

26. An unlikely but illustrative U.S. domestic parallel involving the creation of an unusual and potent blocking coalition may be found in Michael Pertschuk's stewardship of the formerly sleepy Federal Trade Commission (FTC) in the late 1970s. The FTC had recently launched a number of rule-making efforts that directly affected a range of small business interests in the United States, such as funeral homes, used-car dealers, and optometrists. Furthermore, the FTC decided to take on the issue of children's television advertising, which not only threatened major media advertising revenues, but also smacked of First Amendment restrictions. In effect, having energized and unified an enormous coalition of large and small business and media companies—many of whom had been bitter rivals before—the FTC engendered a hail of protest, had its budget and authority slashed, and was even shut down for a while. In part, Pertschuk's unintended legacy was a far more unified and politically effective business community. See Philip B. Heymann, *The Politics of Public Management* (New Haven: Yale University Press, 1987).

27. William R. Moomaw, "A Modest Proposal to Encourage Unilateral Reductions in Greenhouse Gases" (unpublished paper, Tufts University, 1990).

28. Indeed, the legal discussions that led to the Vienna Convention began in 1981, four years after the UNEP had formulated a World Plan of Action on the Ozone Layer. See Thacher, "Alternative Legal and Institutional Approaches," 108–109.

29. Mustafa Tolba, "A Step-by-Step Approach to Protection of the Atmosphere," *International Environmental Affairs* 1, no. 4 (fall 1989): 305.

30. Jessica T. Mathews, "Redefining Security," *Foreign Affairs* 68 (spring 1989): 162–177.

31. Tommy T. B. Koh and Shanmugam Jayakumar, "The Negotiating Process of the Third United Nations Conference on the Law of the Sea," in *United Nations Convention on the Law of the Sea 1982: A Commentary*, ed. M. H. Nordquist (Boston: Martinus Nijhoff, 1985) 29–134.

32. Ibid., 41.

33. See Gardner, *Negotiating Survival*, 24–33.
34. See, generally, Paul R. Ehrlich and Anne H. Ehrlich, *The Population Explosion* (New York: Simon & Schuster, 1990).
35. For a general treatment of the underlying theoretical issues of issue linkage and separation, or "negotiation arithmetic," see James K. Sebenius, "Negotiation Arithmetic: Adding and Subtracting Issues and Parties," *International Organization* 37, no. 1 (autumn 1983): 281–316; or chapter 9 of Lax and Sebenius, *The Manager as Negotiator.*
36. James K. Sebenius, "The Computer as Mediator: Law of the Sea and Beyond," *Journal of Policy Analysis and Management* 1, no. 1 (fall 1990): 77–95.
37. See, for example, Robert L. Rothstein, *Global Bargaining: UNCTAD and the Quest for a New International Economic Order* (Princeton: Princeton University Press, 1979), for an account of an earlier such engagement.
38. World Commission on Environment and Development, *Our Common Future* (Oxford: Oxford University Press, 1987).
39. At present, the OECD countries account for approximately 45 percent of carbon emissions. With the addition of the former Soviet Union and Eastern Europe, the total would rise to 71 percent. See Manne and Richels, "The Costs," 15.
40. Stone, "The Global Warming Crisis."
41. The experience of the Long-Range Transboundary Air Pollution Convention, in which groups of expanding size acceded to the later sulfur and nitrogen oxides protocols, is also generally in accord with this small-scale approach. For a summary, see C. Ian Jackson, "A Tenth Anniversary Review of the ECE Convention on Long-Range Transboundary Air Pollution," *International Environmental Affairs* 2, no. 3 (summer 1990): 217–226.

3 Equal Measures or Fair Burdens: Negotiating Environmental Treaties in an Unequal World

■ EDWARD A. PARSON
AND RICHARD J. ZECKHAUSER
John F. Kennedy School of Government
Harvard University

In negotiations of the FCCC, signed in June 1992 at the Earth Summit, the most contentious issue was whether the treaty would include targets and timetables, that is, quantitative limits on national emissions of carbon dioxide and when they should occur.[1] This dispute represented a substantial departure from the majority of recent experiences in environmental negotiations, in which discussions have routinely focused on the allocation of quantitative national limits on polluting activities. In the climate negotiations, most industrial nations, led by the EC, argued that carbon dioxide targets and timetables are a necessary first step to a more comprehensive and flexible set of obligations. The United States opposed targets and timetables on three grounds: that they are premature given the present level of scientific knowledge about climate change, that early enactment of narrow carbon dioxide targets would bind nations to costly measures that might be hard to reverse even if later found to be unnecessary, and that such targets would obstruct negotiation of a comprehensive long-term management system.[2] The treaty was able to be signed because of a last-minute compromise that characterized national emissions goals in highly ambiguous language, which allowed some governments to assert that the treaty did represent a commitment to specific carbon dioxide targets and others to assert that it did not.[3]

After the Earth Summit, the debate over targets and timetables continued. At a December 1992 meeting in Geneva, some environmental groups and governments argued that parties should begin negotiations immediately on a "CO_2 Protocol" under the framework convention, whose core would be carbon dioxide targets and timetables. On Earth Day in April 1993, President Clinton reversed the previous administration's opposition to targets and timetables and, in carefully measured language, committed to a U.S. limit on greenhouse gas emissions in the year 2000.

This chapter discusses the negotiation of national emissions targets as well as other forms of quantitative national obligations. Questions focus on who will do or give how much, and how will obligations be decided and defined. Other authors, including Mitchell and Chayes in this volume, focus on institutional functions such as monitoring, verification, compliance, enforcement, capacity building, research, and training that precede or accompany the negotiation of control obligations. While acknowledging that these measures are of key importance in nations' achieving and sustaining effective international agreements, we contend that the negotiation, assessment, and comparison of quantitative measures of national performance—emissions targets or equivalent measures—will remain the central problem in negotiations on climate, and indeed in negotiations on any multilateral environmental issue.

We focus on the symmetry of obligations and of interests. We argue that most multilateral environmental treaties impose symmetric obligations on all signatory nations—that is, they require all signatories to do the same thing, in a sense that we make more precise in the next section. But in most cases, the nations joining a treaty are highly asymmetric in their interests on the environmental issue being negotiated—on how strongly they want to improve it and on how much it will cost them to make any particular level of contribution to the effort. While treaties with simple, symmetric obligations offer important negotiating advantages, symmetric deals among asymmetric parties can create both serious inefficiencies and inequities. For environmental issues involving high stakes, such as climate change, comprehensive agreements that are based on simple, symmetric distributions of obligations may be impossible to negotiate. In the final sections of this chapter, we examine the functions that symmetry performs in promoting negotiated agreement. We also develop some preliminary alternative approaches to multilateral environmental negotiation that may provide these functions while more flexibly admitting the disparate interests of asymmetric parties.

Much of our argument is based on an abstract, stylized treatment of international environmental issues, their costs, and measures for their mitigation. We treat international environmental protection as a problem that entails provision for the *public good* among unitary nations, and we treat environmental negotiations as negotiations over what level of the public good is to be provided and by whom. Each negotiating nation is assumed to have a well-defined overall national interest. We also assume that each prefers that the global total of environmentally harmful activities be reduced but that the bulk of the reduction be done by others.

SYMMETRIC OBLIGATIONS IN INTERNATIONAL TREATIES

Most international environmental treaties impose symmetric obligations on all parties—that is, they require all nations signing to do the same thing. That "same thing" may take several forms. It may involve the establishment of common administrative procedures to regulate some activity. The Basel Convention on Transboundary Movement of Hazardous Wastes, for example, requires that national authorities only permit the export of hazardous wastes under certain circumstances and after the importing state has given written notice of its consent to the shipment (UN Environment Programme 1991). Alternatively, treaties may impose common technical standards on some activity, such as loading and ballasting procedures in oil shipping (Mitchell 1993), or they may ban some activity entirely, such as blue whale hunting or ivory importation.

The most significant recent environmental treaties, though, have imposed quantitative national limits on specified emissions or on other forms of environmentally damaging activity. In these treaties, the "same thing" normally means an equal percentage of emissions reductions from a fixed-year baseline, effective for all countries in the same year. Table 3.1 shows several examples from recent treaties and from major proposals and declarations.

In those cases in which treaty obligations do not treat all participants identically, the differences tend to be extremely limited, typically taking the form of a uniform exception for one class of countries. For example, the Montreal Protocol imposed a schedule of phased CFC reductions leading to 50-percent cuts in 1999 for industrial countries, while developing countries were to meet each reduction target ten years later. This ten-year grace period for developing countries has been retained in both subsequent amendments of the treaty, so that industrial countries are now required to eliminate CFCs in 1996 and developing countries in 2006.

Among multilateral environmental treaties, there is only one striking counter-example to the tendency toward symmetric obligations: the EC's Large Combustion Plants directive. This agreement specifies national limits on sulfur and nitrogen emissions that range from 70-percent reductions in Germany, France, the Netherlands, and Belgium to various

TABLE 3.1

Symmetrical Obligations in Selected Environmental Treaties and Declarations

Treaty	Year	No. of countries bound	Measure
Montreal Protocol	1987	139	Cut CFCs 50% from 1986 levels; freeze halons at 1986 levels.
London amendments	1990	93	Eliminate CFCs, halons, and CT by 2000 and MC by 2005.
Copenhagen amendments	1992	34	Advance phaseout dates to 1996 for CFCs, MC, and CT and to 1994 for halons; freeze, then eliminate HCFCs; and freeze methyl bromide at 1991 levels.
Convention on Long-Range Transboundary Air Pollution			
Sulfur Protocol	1985	20	Cut sulfur emissions by 30%.
NO_x Protocol	1988	23	Freeze NO_x at 1987 levels.
"NO_x Club"	1988	12	Cut NO_x emissions by 30%.
VOC Protocol	1991	21	Cut emissions of all volatile organic compounds by 30%.[a]
North Sea (third ministerial conference)	1990	8	50% cuts in 37 chemicals; 70% cuts in dioxin, magnesium, cadmium, and lead.
Baltic Sea (ministerial conference)	1988	6	50% cuts in nutrients, heavy metals, and organic toxins by 1995.
Toronto Declaration	1988	(nonbinding declaration)	20% carbon dioxide cuts by industrial countries.

Notes: CFC = chlorofluorocarbon; CT = carbon tetrachloride; MC = methyl chloroform; HCFC = hydrochlorofluorocarbon; NO_x = nitrogen oxides.

[a] The Volatile Organic Compounds Convention was slightly more flexible than this percentage suggests. While most countries advocated 30-percent cuts on all emissions, a few sought to limit cuts to regions contributing to transboundary emissions, and a few were only willing to accept a freeze. The treaty language permitted signatories to select which of these three sets of obligations they would adopt. Three nations chose a freeze (Bulgaria, Greece, and Hungary), two chose 30-percent cuts in specified regions contributing to transborder fluxes (Canada and Norway), and the remaining fifteen chose overall 30-percent cuts.

smaller reductions in Denmark, Luxembourg, the United Kingdom, Italy, and Spain to *increases* in Greece, Ireland, and Portugal (Commission of the European Communities 1988, Annexes 1 and 2).

ASYMMETRIES OF INTEREST: ORIGINS AND CONSEQUENCES

The prevalence of symmetric measures in environmental treaties is puzzling because the nations participating are often highly asymmetric in their relevant interests. Nations can differ both in the benefits they derive from the environment being improved and in the costs they incur from undertaking measures to improve it. This section considers the kinds of differences among nations that yield such variations in costs and benefits for a particular environmental good and their consequences for symmetric environmental agreements. (Parson and Zeckhauser, 1993, provide a more formal development of the consequences of bargaining among agents with asymmetric interests, using simple two-agent and ten-agent models.)

One nation may value an environmental good more highly than another for several reasons. One nation's citizens may be wealthier, giving a higher valuation of environmental goods.[4] One nation may be more vulnerable to the consequences of a particular form of environmental harm. For example, its soil may be more easily acidified; its coastal regions more extensive and low-lying, hence more sensitive to sea level rise; its economy more dependent on the environmental resource; or its people more sensitive to related health harms. One nation may be larger or more populous, and so capture within its borders a larger fraction of global damage. Alternatively, its citizens or political groups may simply care more about the environment and so be willing to pay more to protect it, perhaps for historical or cultural reasons. These differences may change over time and may be endogenously determined. People may learn to value a good environment by having one, and they may learn to value many kinds of environmental quality, including those not directly observable, by having certain observable kinds. Alternatively, people may grow to accept lower environmental quality by learning to adapt or by simply enduring it.

One nation may also find it more costly than another to restrain activities that contribute to a particular form of environmental harm. Its economy may be more dependent on a particular harmful activity in aggregate or may have more limited opportunities to substitute away from it. This difference may reflect different capital stock structures, natural resource endowments, or levels of technological development.

For some environmental issues, the transmission of harm between countries is asymmetric, as when one country's emissions flow principally into another country. Downwind states import more air pollution from their upwind neighbors than they export to them; downstream riparian states receive pollution from upstream states, but not the reverse. This form of asymmetry does not arise for global-scale environmental issues such as climate change and ozone depletion, but can dominate negotiations over such regional issues as acid deposition, smog, and international river pollution.

Finally, the *means available* to control an environmental harm may impose asymmetries even if the harm itself is borne symmetrically. For example, if the path chosen to reduce emissions is to create property rights in the emissions, then those granted the rights may appropriate the rents that in another regime would have gone to other, former, or potential emitters. With property rights undefined at the outset, as is typical for pollution, then any particular way of defining them is likely to create gains for some and losses for others in a manner that is both relative to the status quo and relative to agents' expectations of the rights they should receive. Examples include the historical fencing of medieval commons and the modern demarcation of ocean fisheries into exclusive economic zones, which has been estimated to represent a $5 billion transfer to the fishing industry in the United States (Christy 1977, cited in Dasgupta 1982: 28). Contemporary discussions over the use of tradeable emissions permits to solve environmental problems, both domestically and internationally, thus focus on the problem of how to define an acceptable distribution of the property rights so created (Joskow 1991; Grubb 1989; Grubb et al. 1992).

These forms of asymmetric interests flow from our simple model of a well-defined national interest that yields national cost-and-benefit calculations for the international environmental issue. A detailed political description of state interests would yield further sources of asymmetric interests. Since the environmental harms, and the costs of reducing national contributions to them, fall on different people within each nation, we also expect nations to be stronger opponents of international environmental protection to the extent that those bearing the emissions-mitigation costs within a nation are more concentrated, better known to each other, and more strongly represented in relevant political bodies than those bearing the costs of the environmental harm to be mitigated.

The existence of these forms of asymmetries of interest is compelling, but some of the asymmetries are hard to observe and measure. There are a few examples of agreements that have been adopted in which the asymmetries of cost were both obvious and large. For example, the 50-percent CFC cuts in the Montreal Protocol applied both to those signatories who had already made 50-percent cuts by eliminating CFCs in aerosol sprays and to those who had done nothing. The compliance costs of the latter group were close to zero, whereas those of the former were substantial. The 1986 moratorium on commercial whaling, although universal in its application, imposed costs only on those nine nations that were still hunting whales (Mitchell 1993). One study found that among the nations that pledged in 1988 to make 30 percent cuts in their emissions of nitrogen oxides, the marginal costs of the agreed cuts ranged from less than DM 300 per metric ton to more than DM 8,000 (Amann 1989; derived from graphs in app. 1).[5] Seventeen European nations agreed in 1985 to make uniform 30-percent cuts in sulfur emissions, but subsequent modeling has shown that transborder emissions transport is so asymmetric that an optimal pattern of reductions would have had some countries reducing by as little as 2 percent, with others reducing by more than 80 percent (Maler 1991, tab. 2).

When nations' interests are highly asymmetric, simple symmetric control measures can bring two kinds of consequences. First, they can create greatly differing marginal costs of control among participating nations, leading to large economic inefficiencies. Consequently, the costs of compliance can be much higher than under a more flexible agreement yielding the same aggregate level of environmental protection.[6] Second, the *total* cost burden imposed by equal-measures treaties can also be highly unequal, suggesting that such agreements may violate plausible standards of fairness.

ADVANTAGES OF SYMMETRIC AGREEMENTS

If equal-measures treaties have such deficiencies, why are they so widespread? First, it is important to note that international bargaining is qualitatively different from domestic bargaining because of the degree to which supervening authority is available, which affects the benefits of symmetric deals. In negotiations over the domestic provision of environmental public goods—for example, in regulatory negotiations to set environmental standards or emissions limits in different regions within a federal state—a federal authority has some power to impose solutions and to enforce trades across issues and over time. Such an authority can overcome the obstacles to agreement that asymmetric interests pose, since a national authority can, at least in principle, impose solutions even over the objections of some affected agents. The solutions imposed can range from benign to malign. For example, with unrestricted side payments, or an ideal distribution of property rights, an authority can impose a range of solutions that might make all agents better off (for representative arguments, see Wirth and Heinz 1991).

Other, less ideal forms of control might leave some or all agents worse off but can still be imposed.[7] With such a supervening authority, complex, asymmetric regulatory solutions are more feasible. It is not surprising that such solutions are more common domestically than internationally. It is also not surprising that the most complex and asymmetric extant international environmental agreement, the Large Combustion Plants directive, was negotiated within the EC, which has extensive institutional linkages and community-level authorities. Even so, reaching this agreement was difficult and slow, requiring five years of intensive twice-weekly negotiating meetings (Grubb 1989: 14).

For international commons problems, sovereign states must bargain in the absence of a supervening authority. Requirements inherent in the process of multilateral bargaining imply several strong advantages to simple equal-measures deals, even when agents' interests differ sharply.

On some issues, asymmetries of prior measures and of concern may partly cancel each other, making equal-measures deals more reasonable than they seem at first glance. For example, if some parties face higher marginal abatement costs because they have taken prior unilateral measures, perhaps hoping to advance the international process by their example, they may in fact be the most concerned and hence the most willing to bear higher costs in a negotiated agreement. Despite an unequal starting point, equal-measures treaties may then be a reasonable basis for agreement, roughly reflecting parties' differing degrees of concern. Such an argument can be made, for example, for the inherent equity of the uniform 50-percent CFC reductions in the 1987 Montreal Protocol.

Such a happy coincidence of asymmetries will not be typical. However, even when asymmetries do not control each other in this way, equal-measures deals still enjoy important advantages. First, they are simple, thus limiting the informational burden on negotiators. Considerations of simplicity grow in importance as the number of participants in a negotiation increases. With N nations participating, an unstructured negotiation over emissions of even one pollutant would require agreement on N distinct national emissions levels, equivalent to the selection of a point in an N-dimensional bargaining space. The number of candidate agreements to consider, and the time required, would grow exponentially with the number of participants. Two or three nations could negotiate with detailed attention to the particular measures each would undertake, but as the number of participants increased, such a negotiation would quickly become impossible. Multiple negotiators need some prior rule, or constraint, to limit the number of agreements under consideration. The presumption that negotiations will concern equal proportional emissions reductions accomplishes this limitation by reducing the bargaining space to a single dimension, requiring agreement only on the common level of reduction that all will undertake.

Second, equal-measures deals restrain the most extreme attempts by negotiators to claim value for their own nations. A prior presumption of equal measures ties reductions in each nation's contribution to reductions in all other nations', so any measure that a participant nation might advocate brings it both benefits (reductions by others) and costs (its own reductions). Whereas any participant in an unconstrained negotiation among highly asymmetric parties could likely find a plausible pretext for contributing a little while others do a lot, the prior presumption of equal measures restrains such fruitless advocacy of extreme positions. All parties may prefer this collective renunciation of opportunistic bargaining.

Third, a principle of equal measures restrains another form of opportunism, the type that arises when a party to a prospective deal threatens to delay or withdraw in order to secure additional advantage. Equal measures avoid such problems by tying together the contributions of different parties in a way that is prominent and simple. Equal cuts represent a *focal point*, an element from a choice set that commands attention by virtue of uniqueness, discreteness, or salience (Schelling 1960). A focal point is defined by negotiators' perceptions, not by objective reality; it need have no particular scientific or economic merit.[8] Because a focal point stands out in negotiators' perceptions, an incremental departure from it, as would be obtained by a slight shading of a negotiated deal to one party's advantage, looks like a big change. Consequently, focal-point solutions in negotiations enjoy a stability that comes from the convergence of negotiators' expectations: each expects others not to try to shade the deal to their advantage, because each expects that to do so itself would risk widespread imitation, effectively reopening the entire negotiation.

In addition, equal-measures deals are easier to ratify because of their salience, simplicity, and resistance to attempts by negotiators to shade the deal. Such deals also have a prima facie appearance of fairness, making it harder for domestic opponents to mount persuasive arguments against them. Paradoxically, an unequal-measures deal may be harder to ratify than an equal-measures one, even for those parties making the smaller of the unequal contributions. Domestic opponents can more easily argue that their country's negotiators should have done better still. This opposition may feed back into the negotiating process itself; the negotiating agent, anticipating this criticism, may be more hesitant to reach agreement.

These advantages of equal-measures deals apply to a single multilateral negotiation. These same advantages may be magnified when the same set of parties expect to negotiate many deals. Even the parties most aggrieved by equal measures in any particular negotiation may be willing to endure them rather than break the norm of equal measures for other negotiations. The presumption of equal measures buys a predictability of negotiations that everyone wants in return for which each may be willing to bear higher costs in some, perhaps even in most, deals. The immediate benefit of breaking the precedent might not be worth the effort needed to negotiate future deals or the ill-will generated, for the consequence would be to risk occasional terrible deals and to endure the high friction and transaction cost of negotiating distributions of obligations from scratch each time.

CLIMATE NEGOTIATIONS: ASYMMETRIES OF INTEREST

We now turn to international negotiations on global climate change, with a particular focus on the problem of negotiating limits on carbon emissions. This section presents information relevant to the negotiation of nations' asymmetries of interest in emissions, abatement costs, and sensitivities; the sections that follow argue against the feasibility of negotiating a climate treaty based on equal proportional emissions reductions and propose alternative approaches to negotiations.

As discussed above, nations may differ greatly in their cost-and-benefit assessments for emissions abatement, reflecting differences in such characteristics as size, wealth, income, tastes, economic structure, and capital stock. The full economic cost a nation bears from abating emissions depends on the particular time path of the emissions constraint imposed, the policies used to achieve it, and a host of characteristics of the national economy (for example, the rates of productivity growth and how these respond to energy constraints, and the ease of substituting fossil with non-fossil energy). To model these characteristics fully would require a dynamic general-equilibrium framework and a great deal of nation-specific data. Such a model would yield a time path of the economic effects of emissions constraint—GNP losses, implicit carbon tax levels, and so forth. Fully characterizing nations' asymmetries of interest in emissions abatement would require consistent application of such a model to many countries. This work has not yet been done, though several sophisticated energy-economic models have been used to study carbon abatement costs in the United States and in a few aggregated world regions.[9] Consistent comparisons of abatement costs between particular nations are only currently available from bottom-up cost estimates that compare available opportunities for technical substitution. These estimates do not permit comparisons of aggregate economic effects, but only of direct marginal costs of substitution. The results of one such estimate for EC countries is presented in figure 3.1, which shows marginal costs per ton of carbon abated by each country relative to a 1988 emissions baseline. The zero points of the curves differ because national emissions are projected to grow at different rates between 1988 and 2010. The variation among countries is very large, and both high and low marginal costs are found among both high- and low-income countries. The countries with the highest costs include Italy, the Netherlands, Spain, and Belgium, while the countries with the lowest costs are Denmark and Portugal.

Less formally, a nation's position in international climate negotiations is also likely to be associated with measures of its aggregate emissions. Many different measures of aggregate emissions may be used in support of negotiating positions, and as Subak and Clark (1990) showed, what measure is used—current or cumulative, national totals or per capita, per hectare of land area, or per dollar of GNP—drastically alters the rank ordering of the largest contributors. Whether all gases or just carbon emissions are counted, and whether sinks are counted as well as sources, also makes for marked shifts in relative accountability, as recent controversy over the accounting system used in the 1990–1991 World Resources

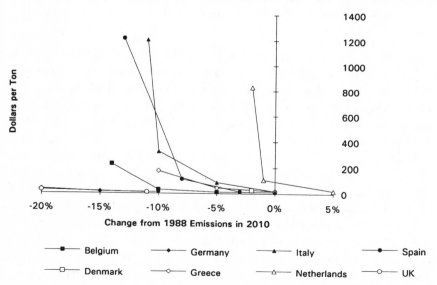

Figure 3.1. National marginal cost curves for carbon abatement, 2010. *Source:* Barrett 1991.

TABLE 3.2

	Variation of Greenhouse Emission Indicators among Specified Nations, 1987	
Location	Carbon emissions (metric tons per capita)	Total greenhouse emissions (kg carbon equivalent per $ GDP)
EC	0.8–3.6	0.17–0.62
OECD	0.8–5.4	0.11–0.62
World	0.008–22.4	0.10–5.6

Sources: World Bank 1992; World Resources Institute 1990; Subak and Clark 1990.

Notes: The EC carbon range excludes Luxembourg, which emits 6.8 metric tonnes per capita; Germany is otherwise highest at 3.6 tonnes per capita. The United States is highest in the OECD at 5.4 tonnes per capita; lowest in the EC and the OECD is Portugal. The highest world per capita carbon emitter is Laos, almost entirely due to deforestation; the other highest emitters include Brazil and several small OPEC states. All emit at around 9 tonnes per capita.

TABLE 3.3

	Variation of Climate Sensitivity Indicators among Nations		
Location	Percent of GNP in agriculture	GNP per capita	Costs per capita of 1-meter sea level rise
EC	2–17%	$4,900–$22,320	$190–$429
OECD	2–17%	$4,900–$32,680	$38–$1,550
World	0–67%	$80–$32,680	$38–$1,800

Sources: World Bank 1992; CIA 1992; Houghton, Jenkins, and Ephraums 1991.

report attests (World Resources Institute 1990; Agarwal and Narain 1991). Large variations in aggregate emissions indicators exist not just between industrial and developing nations, but also among relatively homogenous industrialized nations. Table 3.2 shows ranges of variation for two aggregate emissions measures among countries of the EC, the OECD, and the world.

Currently, no comparative national estimates of damages from climate change exist, principally because there are no reliable biophysical models of regional impacts. The one exception is that estimates are available of costs of nations' protecting against specified rises in sea levels. Still, casual inspection of ranges of wealth, climate, and economic structure among nations suggests that they may differ strongly in their susceptibilities to harm. The first two columns of table 3.3 show ranges of variation on two measures that are plausibly associated with national sensitivity to climate change. It seems logical that countries with much of their economy in agriculture will be more sensitive to change and that those with higher incomes will be better able to adapt to any changes.[10] The third column shows results from one study on relative costs of protecting against a one-meter rise in sea level.

Nordhaus has made systematic, order-of-magnitude estimates of the sensitivity of the American economy to the climate change resulting from steady-state doubled atmospheric carbon dioxide. Estimating production losses by dividing the economy into sectors deemed more and less climate-sensitive, he derived a best guess of 0.25-percent steady-state GNP loss and conjectured an upper bound of 2 percent, implying a marginal damage per ton of carbon emissions of $0.57 to $13 (with an upper bound of $4.50 to $106). We have found no comparable analysis for other countries.

The problem of finding plausible quantitative estimates for national sensitivity to climate change is so great that other studies have inferred damage functions from abatement costs by making some assumption of optimal behavior. For example, Barrett's (1991) recent work on the EC assumed that the EC's pledge to freeze carbon represents the optimal emissions level, and hence derives marginal damage of $75 per ton of carbon emitted. This approach follows Maler's (1991) study of European acid rain, which assumed that nations in 1984 emitted at levels that maximized their welfare taking other nations' emissions as given (that is, a Nash equilibrium), and thereby derived marginal damages from known marginal abatement costs.

Although cross-national comparison estimates of costs and benefits relevant to climate change are suspect, weak on the abatement side and even weaker on the sensitivity side, there are plausible grounds to suspect that international variations in these costs are large.

CLIMATE NEGOTIATIONS: EQUAL CUTS AND OTHER FOCAL POINTS

Many multilateral environmental treaties have settled on equal-emissions cuts despite substantial asymmetries of interest among the negotiating nations. In addition to the general factors favoring the negotiations for equal reductions discussed above, these negotiations may also have been supported by the perceived need to achieve agreement quickly. Many observers of climate negotiations argue that a similar deal should be made soon on this issue as well by setting a freeze or requiring equal percentage reductions in national carbon or greenhouse emissions for industrial countries, plus a set of later or smaller emissions-control obligations for developing countries, and by establishing a fund to support them. They cite the Montreal Protocol in particular as a precedent for an equal-measures treaty on carbon emissions.

So why fuss about symmetric or asymmetric reduction obligations? If international environmental diplomacy has always happily pounded asymmetric pegs into symmetric holes and accepted the resultant inefficiency and inequity, why not continue to do so? We contend that for climate change, this traditional approach is unlikely to work. A climate treaty based on equal measures is likely to be impossible to negotiate and unsustainable if negotiated. The principal reason for this failure is that the costs on both sides of the ledger—the abatement costs of a serious attempt to slow the rate of climate change and the impacts of plausible rates of change if emissions are not restricted—may be much larger than those of any environmental agreement to date. The inefficiencies and inequities inherent in across-the-board reductions, which can reasonably be expected to vary with the total costs, are consequently also likely to be much larger than for any environmental issue yet dealt with. It is our view that these costs are liable to be sufficiently large to obstruct agreement on the kinds of simple formulas that have been pursued so far.

We particularly dispute the relevance of the successful negotiation of across-the-board CFC cuts in the Montreal Protocol. In addition to the fact that total abatement costs for CFCs were orders of magnitude smaller than they would be for serious carbon-emissions reductions, agreement on this simple formula was facilitated by four factors not present for climate change.[11] First, CFC negotiators worked under a high level of perceived urgency, because there was substantial evidence that the environmental threat was worsening as they negotiated. With such great perceived need for rapid agreement, the simplicity and salience benefits of equal cuts took on even greater importance. For climate, while surprises are of course possible, most observers expect the system to evolve over a space of decades or centuries; therefore, climate change will never have the same perceived urgency for immediate action. Second, the costs of abatement were narrowly borne by a few internationally traded industrial sectors, so the concerns of potential opponents focused more on the competitiveness effects of relative cost burdens than on absolute costs. Third, because the number of parties whose agreement was required to strike a viable deal was relatively small, it was possible to accommodate the most serious concerns of a few parties by minor variation in the form of controls. Finally, it became clear by the time of the 1990 amendments that the desirable action on ozone-depleting chemicals was the most simple, salient, and equitable solution of all: a complete phaseout.

If we are correct that climate negotiations based on across-the-board emissions reductions are infeasible, what approach do we propose? We argued above that a presumption of equal reductions facilitates negotiations in three ways: (1) by simplifying the set of proposed solutions to be considered through a reduction in the dimensionality of the bargaining space, (2) by restraining the extremes of opportunism by coupling increases in each nation's contribution to increases in all others', and (3) by reducing incremental attempts to shade the agreement by providing a focal point that stands out conspicuously and discretely from other nearby solutions. However, equal reductions are not the only way to realize these advantages; many other principles that tie together contributions of various parties to a bargain could serve the same functions. The following is an illustrative list of principles that meet these conditions. It is by no means exhaustive.

- Fixed equal targets for emissions reductions (proportional or absolute)

- Equal reductions per person (proportional or absolute)

- Equal reductions per dollar of GNP (proportional or absolute)

- Equal annual incremental changes (in emissions per person, emissions per dollar of GNP, energy intensity, or other measures)

- Equal entitlements to emit per person

- Equal entitlements to emit per dollar of GNP

- Equal entitlements to emit per unit of land area

- Reductions in emissions to incur equal abatement cost per person

- Reductions in emissions to incur equal abatement cost per dollar of GNP

- Allocation of equal fractions of governmental budgets to emissions-abatement programs

- Taxation of emissions at equal rates

- Taxation of specified activities related to emissions at equal rates (specific or ad valorem)

- Equalization of marginal abatement costs (proximate or economy-wide) through nontax measures

- Adoption of equivalent specific technical regulations or goals (for example, all new power plant investment to meet specified efficiency goals or auto fuel-efficiency standards)

- Revision of any of these principles to reflect cumulative emissions history (a simple integral or a discounted integral)

The observation that many principles can serve the same simplicity and salience functions as equal percentage reductions could improve the prospects for reaching agreement. In any particular negotiation, some of the plausible candidate principles will be less severe in their inefficiencies and inequities, and hence less obstructed, than others. But even such an expanded set of candidate principles still leaves two serious obstacles to the negotiation of an agreement: one a consequence of the expanded set of candidate principles, the other a difficulty that the expanded set shares with equal cuts.

First, how will one principle to guide negotiations be chosen from many candidates? In some past negotiations, it seems that a cut by equal percentages was the only approach ever considered. However, when negotiators realize they must choose among plausible competing principles, this very recognition that a choice must be made ensures that no principle can enjoy the same prominence that the option for equal reductions does when it alone is on the bargaining table.

Moreover, in any particular negotiation, some of the candidate principles will naturally and visibly favor certain parties. With both principled and interested differences over what rule should be followed, there will be ample opportunity for each negotiator's cognitive biases to make the most advantageous rule seem coincidentally to be the fairest and most reasonable. When evaluating any particular proposed deal, each negotiator is likely to perceive most vividly the principles according to which this deal leaves his or her nation overburdened and others underburdened. It is not sufficient merely to assert that negotiators will be disingenuous, though they will be. Rather, sincere and opportunistic disagreements are liable to mingle in the space that cognitive biases allow, making the negotiation far more contentious and difficult than it would be under either simple principled disagreement or pure conflict of interest.[12] With many discrete, plausible principles available to distribute obligations, choosing one will require a negotiation of great subtlety and difficulty.

Second, any one of these principles is liable to suffer from one of the same difficulties as equal cuts. In informational terms, the essential characteristic of a proposed principle is the number of degrees of freedom it leaves open for negotiation. An appropriate principle for any particular negotiation will leave enough to allow the negotiations to accommodate the relevant range of interests and special cases, but few enough to be manageable. Much of the difficulty of equal cuts is that they leave just one degree of freedom to negotiate, but this is also the case for most of the principles on the list above *if they are applied strictly*. If equal cuts are not applied strictly, but nuanced with special exceptions and interpretations, particularly in negotiations among a large number of parties, the advantages of simplicity and salience are liable to be at least partially lost.

Four criteria for desirable principles suggest themselves. First, principles should not imply too grievous a departure from efficiency. Second, they should be anonymous, in that a nation's treatment should not depend directly on its identity. Third, they should not leave any nation much worse off than with no agreement. Collective pressure to join an agreement may overcome some disadvantage to a particular nation, but the disadvantage would have to be modest. Fourth, principles should have a logically defensible claim to fairness. In a negotiation on a complex set of policies, however, claims of fairness may be difficult to establish and may bear differently on different aspects of the policy. For example, if some activity (for example, coal burning) is agreed to be strongly identified with the environmental harm to be controlled, there may be quick agreement that a tax on coal is a fair component of a solution. But the question of what ways are fair to distribute the resultant surplus is likely to be much more difficult.

OTHER APPROACHES TO CLIMATE NEGOTIATION

The foregoing discussion suggests a problem: when the stakes are potentially enormous and parties are asymmetrically situated, equal-cut deals are likely to be blocked. Other principles can meet the same negotiating needs, but any one is also likely to be highly limited, and negotiators' choosing any single principle explicitly is liable to be contentious, difficult, and time-consuming. This section suggests some ways of structuring negotiations that offer the potential of moving beyond these difficulties.

NESTED NEGOTIATIONS

Many of the difficulties of negotiation posed by equal cuts and other simple principles are consequences of the large number of nations participating in negotiations. To the extent that the number of effective participants in a negotiation can be reduced, the difficulties are mitigated. While Brenton (1992) proposed that climate negotiations be restricted to those twenty or so nations that are big, rich, or influential enough to matter, other approaches could achieve the same simplification without excluding nations.

The most widely used of these approaches is to group participating nations into a few classes in which members are similarly situated, to agree that all members of a class will do the same thing, and to negotiate over what that "same thing" will be for each class. If the set of all negotiating nations is too diverse for all to take the same measures, but if they fall into a few clear, politically salient, internally homogeneous categories, then this approach could yield more easily negotiable and efficient agreements. The most commonly used categories, of course, are industrialized and developing countries; separate obligations for these groups have been enshrined in the Montreal Protocol, as well as in the recent negotiations on climate change and biodiversity. However, even this simple division into two classes of obligations has been surprisingly limited in application. The most detailed application so far has been to grant developing countries a grace period and financial assistance in meeting precisely the same obligations as industrial countries rather than to negotiate a separate set of obligations.

Even if pursued more aggressively, the approach of grouping countries into classes has two limitations. First, even nations within seemingly homogeneous classes may be sufficiently dissimilar that imposing uniform measures on them is excessively costly or unfair. The abatement costs shown in figure 3.1 suggest that even within the EC the gains from relaxing a constraint of equal reductions are great. Second, criteria defining classes may be sufficiently unclear that it will be difficult to decide where to assign border cases and to resist pressure to increase the number of categories. There is, for example, a legitimate case for the separate treatment of economies in transition from central planning to markets, as has been proposed in both CFC and climate negotiations. In the climate negotiations, both low-lying coastal and island nations and African nations subject to desertification also claimed special status because of their high climatic vulnerability. The difficulty of defining agreed categories and who belongs within them can be mitigated in part by the use of ranges of relevant numerical measures of responsibility, ability to contribute, or sensitivity. The Montreal Protocol, for example, applies the criterion that national CFC consumption be less than 0.3 kilograms per capita for the treatment of developing countries.[13]

One way of enlarging the number of classes of countries receiving different treatment is for informal groups of activist nations to self-select to agree to strong measures. On two occasions (at the 1988 NO_x Protocol and at the 1990 amendments to the Montreal Protocol) such informal "clubs" of like-minded countries have announced themselves at the conclusion of a negotiation that has failed to yield action as strong as they had advocated. On one other occasion, a club formed before a treaty was negotiated to increase the pressure for a treaty: the "30 Percent Club" on controls for sulfur emissions. One proposal would enshrine this practice by having parties agree that reduction programs undertaken after a fixed date would count toward a nation's obligations in future treaties (Moomaw 1990). In terms of promoting future actions, however, such a proposal cuts both ways. It would make unilateral leadership easier, since early leaders would collect later benefits; but it would reduce the effectiveness of such leadership, since early leaders would be in a weaker position to induce others to follow their lead.

A distinct approach to the reduction of the effective number of negotiating parties would be to structure global negotiations as deliberations among a few large regional blocs that are to take on regional emissions-control obligations. One can imagine a negotiation over the distribution of *regional total* emissions-control obligations among the EC nations, an expanded list of countries agreeing to the North American Free Trade Agreement, an East-Asian trading bloc, a coalition of nations from Eastern Europe and the former Soviet Union, and perhaps a couple of major developing countries. Each bloc would then face a subsequent internal negotiation over how to meet its aggregate obligation. Unlike the approach based on distinct *national* obligations for nations within each class, this approach permits the efficiency gains available from a distribution of obligations within each bloc. It would also permit diversity of implementation approaches between blocs. This approach is of course highly speculative and would depend on the continued development of strong regional economic organizations with increasingly rich institutional capacity to facilitate intraregional trades and redistribution. That the EC ultimately agreed on national emissions quotas in its Large Combustion directive provides a modestly hopeful example for such an approach; that the same nations abandoned their attempt to move from a pledged EC-wide carbon freeze to an agreement on national emissions limits after one year's work provides a cautionary counter-example of how difficult such negotiations can be.

FORMULAS

A second broad approach, which might help to overcome the difficulties of having a multiplicity of plausible principles, would be to negotiate a formula that defines national emissions entitlements, or changes from the status quo, on the basis of national characteristics such as population, GNP, current emissions, or factors plausibly associated with national responsibility, sensitivity, or need for various emitting activities. In negotiating the factors to be included in a formula and their coefficients, parties would, in effect, negotiate a weighted average of several plausible principles for the allocation of contributions, implicitly blending considerations of efficiency with various conceptions of equity.[14]

A similar proposal was advanced in CFC negotiations in 1986, under which national CFC quotas would be linear functions of population and GNP, with the relative shares to be negotiated (Environment Canada 1986). It is instructive that the proposal was dropped in favor of equal cuts because its quotas departed too far from the status quo. Any significant weighting of population would have granted vast overentitlements to developing countries, while even the GNP factor failed to track closely enough the variation in use among industrial countries.[15] However, such a scheme may be more feasible for climate, at least relative to the alternatives. While Nitze's proposal, in which carbon emissions quotas are distributed 50 percent according to population and 50 percent according to present emissions, is no doubt too simplistic, a negotiated formula using three to five factors may be more viable. Such an approach may be particularly viable as the basis for the initial distribution of allocations within an emissions-trading system, since subsequent trading would tend to reduce the remaining inefficiencies in the allocation. The strongest precedent for this approach is the system of defining quotas used in the IMF, which employs a formula based on national income, reserves, exports, imports, and the variability of exports. The coefficients of this formula have been renegotiated twice in the IMF's history (Lister 1984). An approach to negotiations based on a formula could be combined with the division of countries into classes, with a different formula used for each class.

In complex negotiations that seek to include many forms of national policy and action, formulas can play a second role. They can be used to express equivalence between different forms of national action, reducing many distinct obligations to a single aggregate obligation. For example, common metrics can be used to express the relevant environmental equivalence of different kinds of emissions. This has been accomplished with the adoption of Ozone-Depletion Potentials to express the effect of different CFCs in a common, environmentally relevant metric. The pursuit of equivalent "Global Warming Potentials" to weight emissions of various greenhouse gases has proved problematic so far.[16] A comprehensive approach to emissions of all gases, as advocated strongly by U.S. negotiators, among others, will require some such scheme for trading off emissions of various gases.

This approach could be extended to include negotiated trade-offs between other forms of treaty-relevant action. Relationships could be negotiated between emissions reductions achieved sooner and those achieved later (equivalent to a negotiated discount rate), between reductions achieved through domestic action and through funding of action abroad (which is akin to the offset factor used in domestic emissions trading and which allows a balancing of the views of trading advocates and those who contend that each party must accomplish some of its own reductions), or between the limitation of emissions and the management of sinks.

TREND OBLIGATIONS

An alternative to defining obligations as fixed endpoints to be reached by a specified date is to define them as trends, that is, rates of change from status quo or past trend lines that must be achieved each year. Progressive obligations could be defined in terms of targets such as trend lines in national emissions or in terms of energy or carbon intensity of the economy. Alternatively, they could be defined in terms of policy inputs—progressively increased tax levels or expenditures on emissions control. Because energy sources that contribute to greenhouse emissions are subsidized in many nations, negotiators could do as the GATT has done with tariffs: announce their current fossil fuel subsidies, then agree to progressively phase them out over time. A schedule of constant absolute reduction or constant proportional reduction would be feasible (Finger and Olechowski 1987), perhaps modified by provisions to calibrate the phaseout in proportion to the carbon intensity of the fuels subsidized. Unlike tariff reductions, such schemes need not stop at zero, but could continue toward progressive taxation of formerly subsidized activities. Such an agreement would, of course, have a finite lifetime and be subject to successive rounds of renegotiation as the policies phase in.

This type of agreement simply provides a shift in focus from fixed-year goals to annual incremental ones. From a negotiating standpoint, this change in view offers several important advantages. First, progressive obligations integrate the understanding that no one knows how tight emissions limits should be, hence respecting nations' concerns about setting a precedent that might be wrong. Second, progressive obligations integrate the observation that since little is known about the effectiveness of emissions-control programs, nations are truly unable to pledge to meet a fixed target. By avoiding fixed targets (for good reasons), this approach also denies parties the obstructive rhetorical device of criticizing a proposed form of obligations as if they will extend unchanged forever.

Since attainment of targets over a period as short as a year is for the most part a consequence of random fluctuations, this approach basically amounts to establishing a continuing negotiation over the appropriateness and effectiveness of national emissions-reduction programs. The essence of this approach lies in its review process—both periodic negotiated review of the appropriateness of the annual incremental targets and review of national measures—which should strive both to give national policymakers the right incentives (to try hard, be flexible, and report progress honestly) and to maximize learning through international comparison of experiences.[17] Such a system presupposes a very different forum from broad, comprehensive, diplomatic negotiations. Rather, this system seems to require a small, functional, professional group that would develop good working relations as its members continually negotiate, thus acquiring both skill at knowing what works and the ability to consult effectively enough with national officials to apply the pressure needed to keep plans effective and targets honest.

SIDE PAYMENTS

It is a common, and correct, observation that one of the most effective ways of improving the negotiation of environmental agreements would be to decouple the decisions of where and how emissions are reduced from the decisions of who pays for the reductions. This decoupling is necessary to approach minimum-cost allocations of reductions, which require that marginal costs of abatement be equalized across different sources, nations, and regions. Achieving efficient solutions based on marginal costs of abatement does not depend on precise damage estimates; even if such estimates remain elusive, any agreed world limit on emissions can be achieved at least cost by an equalization of marginal abatement costs. Depending on the state of knowledge concerning abatement costs, and how costs change over time, the principle of equalizing marginal abatement costs can come close to specifying a unique desirable configuration of national emissions targets.

The principle of equalizing marginal abatement costs has, however, had essentially no influence in any environmental negotiation to date. While this may partly reflect dispute and ignorance about how large the marginal costs will be and how to measure them, it is more likely a consequence of the severe distributive implications of the principle. Most observers agree that abatement costs are generally lower in developing countries than in industrial countries, because present capital stock is so inefficient and present emissions-control efforts so limited that much cheap improvement is obtainable and because much of the impact on future emissions can be realized in new investment rather than in retrofits. The principle of equalizing marginal abatement costs would consequently specify larger abatement in developing than in industrial countries. As argued above, there are also likely to be large disparities in marginal abatement costs among seemingly similar industrial countries and among developing countries. The consequence is that, absent other policy measures to redistribute costs, the costs of a system based on equalizing marginal costs will be borne highly unequally and are liable to include a large implicit transfer from developing countries to industrial ones.

Since these distributional implications are widely viewed as morally objectionable, and are liable to block negotiated agreement, an allocation of emissions reductions that even approaches efficiency is only possible if it is accompanied by parallel measures to redistribute the burden of costs. This is the side-payment debate. An efficient or nearly efficient solution requires the allocation of emissions on the basis of marginal abatement costs and a completely separate negotiation to allocate costs.

The reality of environmental negotiations is far from this ideal. First, international side payments seem to be hard to negotiate and hard to deliver. In 1990 it was surprisingly difficult to obtain U.S. consent to an international obligation of only $13 million per year for three years to support international CFC reductions. The United States chronically fails to meet its long-established obligations to the UN and the development banks, even for such a widely supported activity as UN peacekeeping operations. Even more problematic than international payments to developing countries, for which some domestic constituency exists, would be side payments to other industrial countries—such as European countries or states of the former Soviet Union. Moreover, it seems that the explicit exchange of money among nations is even harder politically than is mutual accommodation on substantive issues intended to achieve the same result. It is ironic that the strongest example of a complex, asymmetric international environmental agreement, the EC's Large Combustion Plants directive described above, appears to use asymmetric emissions targets to achieve distributive goals at the cost of efficiency losses rather than use the distribution of targets to reduce costs with compensating financial transfers. The largest reductions are required of countries whose marginal costs are probably the highest.

The question of compensation seems to have been effectively solved in the Montreal Protocol negotiations. In the 1990 amendments, the first negotiation with forceful developing-country participation, major developing countries agreed to join the industrial countries in CFC phaseouts if their full marginal costs of doing so were compensated. This compensation was achieved by raising a multilateral fund from the industrial-country parties and by creating a small executive committee to administer expenditures from the fund according to specific negotiated criteria.[18]

However, climate change negotiations will raise the question of transfer payments, in particular the question of how large they should be, in a far more difficult form. An agreement such as the Montreal Protocol, which pays precisely the developing countries' marginal costs of compliance, leaves the recipients no better off than without the agreement. In principle, it leaves them indifferent between joining and not joining the treaty (neglecting for the moment that the treaty's penalties against non-parties make not joining worse than the status quo). If the negotiated level of controls is even roughly right, the emissions controls to be realized in any country should be worth more to the rest of the world than they cost the country to implement, possibly much more. Unless the investment or project to control emissions brings external benefits to the country undertaking it (which may be the case), for the donors to pay only incremental cost is to award them the entire surplus realized in the transaction.

There is substantial evidence in the recent negotiating record that representatives from developing countries perceive their bargaining position to be much stronger than would warrant such a disadvantageous outcome. If the industrial world's concern about greenhouse emissions remains at present levels or grows substantially, then the range of different development plans of such nations as China, India, and Brazil may include some that impose substantial harm on the rest of the world. For them, this possibility can represent a threat, which can be used to extract transfers larger than the costs they would incur in changing paths. Such demands can be characterized morally in terms ranging from compensation for historical inequity and exploitation to claiming a share of a globally created surplus to blackmail.[19]

The redistributive negotiations related to climate change have not yet been seriously engaged. Financial negotiations represented the most divisive item on the Earth Summit agenda, but the resolution was a standoff: developing countries accepted minimal new substantive environmental obligations, and industrial countries accepted minimal new financial obligations.[20]

As the redistributive negotiation is joined more seriously, it is likely to be more divisive as its distributive component becomes more explicit. Consequently, there may be large negotiating benefits to schemes that blend the negotiation of emissions obligations with the negotiation of transfers, in particular schemes that make use of marketable emissions permits or that distribute emissions obligations to multination regional blocs, which then subsequently engage in internal distributive negotiation. The example of the EC suggests that a general increase in the tightness of institutional linkages among nations would also facilitate such negotiations, allowing an increasingly flexible system of compensation through trading across issues without cash changing hands. The example of the Montreal Protocol suggests that there may be a benefit to systems that partly rely on delegating distributive decisions to small, professional institutions.

CONCLUSION

International negotiations over climate change will span years or decades, with many successive generations of agreements. The scope of the job to be achieved and the time span both suggest that the ultimate international management of the issue will involve major institutional innovations. Whatever pattern of obligations is adopted, with the accompanying review provisions and institutional measures, will have to function in a world in which huge changes are taking place. The number and character of the states making up this world will vary, as will the intensity of interstate linkages and conflicts. The most confident prediction we can make is that the ultimate outcome is highly unpredictable.

We had three goals in this chapter: first, to understand the nature of past environmental agreements, and in particular their extraordinary focus on equal contributions; second, to appreciate fully the salience and appeal of the equal proportional reductions approach to climate negotiations in order to analyze its pitfalls and weaknesses; and third, to initiate the search for principles for designing effective and efficient agreements for controlling climate change. Any agreement, we believe, must rely on some formula that sharply limits the dimensionality of the problem. The equal-reductions formula, the standard of the past, will not work. However, we conjecture that potential signatory nations will be willing to abandon this precedent when doing so facilitates reaching more and significantly better agreements. Even prodigious efforts by scholars and negotiators to identify the principles that must underlie workable treaties on climate change control is certainly of merit, since the stakes involved are potentially enormous.

ACKNOWLEDGMENTS

This chapter was partially supported by the U.S. EPA through the Harvard Global Environmental Change Project and by the Decision, Risk, and Management Science Division of the National Science Foundation, whose support is gratefully acknowledged. We thank Herman Cesar,

Henry Lee, Howard Raiffa, James K. Sebenius, Robert Stavins, Larry Susskind, David Victor, and Philip Zelikow for comments and criticism on drafts of this chapter.

NOTES

1. Carbon dioxide, which is emitted from fossil fuel combustion and land-use change, is the largest anthropogenic contribution to climate change, representing about half of total radiative forcing. Other major contributions include methane, nitrous oxide, CFCs and related chemicals, and tropospheric ozone.

2. Detailed summaries of the negotiations are available in *ECO*, a daily newsletter published electronically by journalists and environmental organizations that observed the negotiating sessions, which is available on the electronic bulletin board service Econet. The U.S. position was forcefully criticized by then Senator Albert Gore in a best-selling book (Gore 1992). A widely accepted international consensus of current scientific understanding of climatic change is presented in Houghton, Jenkins, and Ephrams (1991) and IPCC (1992).

3. In addition to the ambiguous language on emissions targets, the FCCC contains extensive institutional measures. It requires nations to prepare consistent inventories of national emissions, develop and report on national plans to limit emissions, and support and coordinate relevant research. It also creates a secretariat and two subsidiary bodies, one to support implementation of the treaty and one to bring relevant new scientific information into future negotiations.

4. Higher values for environmental harm are most likely to be positively correlated with wealth if the harm manifests itself principally in the form of health risks or a loss of consumption. However, if the dominant effect is loss of productivity, the relationship to wealth will depend on relative effects on marginal productivities of different factors of production. Historical experience in a broad range of countries supports this theory in suggesting that environmental quality is unlikely to be an inferior good, that is, one which is demanded less as income rises.

5. The marginal cost of a particular level of emissions reduction is the extra cost imposed by the last unit reduced.

6. Economists have widely attacked equal-measures deals for these inefficiencies. For example, a recent World Bank working paper criticized the Montreal Protocol on the grounds that equal proportional

reductions are inefficient given different control costs and that the developing countries' ten-year grace period represents an in-kind transfer and is hence inefficient relative to a cash transfer (Bohm 1990). Of course, given that direct side payments among nations are often infeasible, such agreements may be economically inefficient; it may be impossible to redistribute *emissions* so as to make somebody better off and nobody worse off.

7. Sometimes, controlling the exploitation of a commons is accompanied by its expropriation. If former users are excluded without compensation, clearly their lot is worsened. Former users are also worse off when exploitation is controlled through a tax whose proceeds are not redistributed (Weitzman 1974; Rolph 1983, cited in Ostrom 1990: 215).

8. Typical kinds of focal-point solutions could include negotiators setting some salient quantity equal for all parties, setting some salient quantity to zero or another round number (uniform reductions are most often by 10, 30, 50, or 90 percent), or keeping some salient quantity at its status quo value.

9. For example, Jorgenson and Wilcoxen (1991) used a dynamic general-equilibrium model of the U.S. economy to model the impacts of a freeze and a 20-percent cut in carbon emissions; they showed annual GNP losses ranging from 0.3 to 1.6 percent and required carbon taxes from $9 to $60 per ton. In their contribution to this volume, Jorgenson and Wilcoxen use the same model to project the time path of a carbon tax necessary to hold U.S. carbon emissions at 1990 levels in perpetuity. Manne and Richels (1990, 1991) applied a somewhat simpler model, linking a moderately detailed energy-sector submodel to a macroeconomic model, both to the United States and to a five-region world, and found substantially larger GNP losses and required carbon taxes: up to a 3-percent loss in GNP for the United States and a $250 per ton carbon tax. Edmonds et al. (1993) have initiated modeling of the impacts of several forms of limits on carbon emissions in a nine-region world.

10. The few existing analyses of projected impacts on climate change place agriculture among the most climate-sensitive sectors (for example, Nordhaus 1990). Ausubel (1991) argued that rich industrial societies already control their climates, suggesting that sensitivity declines with increasing national income. A rich country, however, may be less sensitive in relation to GNP but still willing to pay more in absolute terms to avoid a change.

11. CFC reduction cost estimates also dropped sharply over time. Large

cost reductions normally arise in the abatement of any pollutant—cheaper ways are found to control once solutions begin to be sought—but it is highly unlikely that future reductions in greenhouse abatement cost estimates will be reduced to the magnitude of CFC control costs.

12. Messick and Sentis (1983) presented relevant experimental data from individual behavior. People were presented with choices (dividing a sum of money between themselves and another, absent person) in which they could act completely opportunistically with no penalty (keep all the money), but to which several obvious principles of fairness could also be applied (equal division, division according to hours worked, etc.). They tended strongly to choose according to those principles that most favored them, but rarely chose to keep all the money.

13. This objective criterion has not spared the parties' contention over who gets the status. The 1991 meeting of the parties had to deal with attempts by Turkey to be added to the list and by Jordan and Malaysia to remain on it despite large apparent increases in their CFC consumption.

14. In principle, such a formula would not need to weight the contributing principles linearly, although each departure from straightforward linear weighting would further sacrifice the advantages of simplicity.

15. Distribution of quotas according to GNP, for example, would have bestowed large surpluses on the Scandinavian countries and the former Soviet Union (Parson 1993).

16. Adoption of the Ozone-Depletion Potentials standard has been credited with breaking a crucial deadlock in the CFC negotiations in that it allowed Japan to increase use of CFC-113 by crediting against it reductions achieved in CFC-11 and CFC-12 (Benedick 1991). The search for consensus Global Warming Potentials remains problematic, however, because the contributions of different gases depend strongly on the time horizon used. Selecting a particular time horizon, which is equivalent to deciding on the relative valuation of present and future effects, inevitably weights the relative assessed contributions of different source gases and, consequently, the total contribution of different countries. Short horizons stress the contribution of short-lived gases such as methane, while long horizons stress carbon dioxide. There is no scientific basis for the establishment of one time horizon over another as the correct one (Houghton, Jenkins, and

Ephraums 1990; IPCC 1992; U.S. Task Force on a Comprehensive Approach to Climate Change 1991).

17. This approach has points in common with the "pledge and review" scheme advanced in the climate convention negotiations between Japan and the United Kingdom in 1991 (although that system did not include specific targets) with Schelling's (1992) discussion of the procedures used to oversee disbursement of Marshall Plan money, and with Victor's (1991) proposal for a general agreement on climate change.

18. Once the principle of funding incremental cost was negotiated, the detailed work was delegated to a committee of fourteen countries (seven industrial and seven developing), which work with multilateral development agencies to develop specific investment plans for each developing-country party. The process shows promise because the relatively small group of national representatives involved have built good working relationships and because their fairly specific mandate and fixed budget have made the group an unsuitable forum for the large and contentious distributive questions.

19. Leverage by developing countries in the climate issue is discussed in MacNeill, Winsemius, and Takushiji (1991).

20. The language on development assistance was crafted artfully to avoid representing any new commitments. The climate and biodiversity conventions did, however, include provision for industrial countries to fund the cost of those monitoring and reporting responsibilities that the developing countries accepted (Haas, Levy, and Parson 1992).

REFERENCES

Agarwal, Anil, and Sunita Narain. 1991. "Global Warming in an Unequal World: A Case of Environmental Colonialism." New Delhi: Center for Science and Environment, January.

Amann, Markus. 1989. "Potential and Costs for Control of NO_x Emissions in Europe." Working paper SR-89-1. Laxenburg, Austria: International Institute for Applied Systems Analysis.

Ausubel, Jesse. 1991. "Does Climate Still Matter?" *Nature* 350 (25 April): 6320.

Barrett, Scott. 1989. "On the Nature and Significance of International Environmental Agreements." London Business School, May. Mimeo.

———. 1991. "Reaching a CO_2 Emission Limitation Agreement for the

Community: Implications for Equity and Cost-Effectiveness." Brussels: Directorate-General for Economic and Financial Affairs, Commission of the European Communities, September.

Benedick, Richard E. 1991. *Ozone Diplomacy: New Directions in Safeguarding the Planet.* Cambridge: Harvard University Press.

Brenton, Anthony. 1992. "Machiavelli in the Rainforest." Seminar presentation, Harvard University, Harvard Seminar on International Environmental Institutions, 4 December.

Bresnahan, Timothy. 1981. "Duopoly Models with Consistent Conjectures." *American Economic Review* 71, no. 5 (December 1981): 934–945.

Bohm, Peter. 1990. "Efficiency Issues and the Montreal Protocol on CFCs." Environment working paper 40, World Bank, Washington, DC, September.

Carraro, Carlo, and Domenico Siniscalco. 1991. "Strategies for the International Protection of the Environment." Paper presented at National Bureau for Economic Research Summer Session on the Environment, Cambridge, MA, August 1991. Mimeo.

Commission of the European Communities. 1988. "Council Directive on the Limitation of Emissions of Pollutants into the Air from Large Combustion Plants." Brussels, Working document ENV/88/19. 22 June.

Cornes, Richard, and Todd Sandler. 1983. "On Commons and Tragedies." *American Economic Review* 73, no. 4 (September): 787–792.

Dasgupta, Partha. 1982. *The Control of Resources.* Oxford: Blackwell.

d'Aspremont, Claude, A. Jacquemin, J. J. Gabszewica, and J. Weymark. 1983. "On the Stability of Collusive Price Leadership." *Canadian Journal of Economics* 16, no. 1: 17–25.

Donsimoni, M.-P., N. S. Economides, and H. M. Polemarchakis. 1986. "Stable Cartels." *International Economic Review* 27, no. 2 (June): 317–327.

Edmonds, James E., David Barns, Marshall Wise, and My Ton. 1993. "Carbon Coalitions." Unpublished paper, Global Environmental Change Program, Pacific Northwest Laboratories, Washington, DC.

Environment Canada. 1986. "A Canadian Contribution to the Consideration of Strategies for Protecting the Ozone Layer." Paper presented to an EPA workshop, Leesburg, VA, September.

Finger, J. Michael, and Andrzej Olechowski, eds. 1987. *The Uruguay Round: A Handbook on the Multilateral Trade Negotiations.* Washington, DC: World Bank.

Gore, Albert. 1992. *Earth in the Balance: Ecology and the Human Spirit.* New York: Houghton Mifflin.

Grubb, Michael. 1989. *The Greenhouse Effect: Negotiating Targets.* London: Royal Institute of International Affairs.

Grubb, Michael, J. Sebenius, A. Magalhaes, and S. Subak. 1992. "Sharing the Burden." In *Confronting Climate Change: Risks, Implications, and Responses,* ed. I. M. Mintzer. Cambridge: Cambridge University Press.

Guttman, Joel M. 1978. "Understanding Collective Action: Matching Behavior." *American Economic Review* 68, no. 2 (May): 251–255.

Haas, Peter M., Marc Levy, and Edward A. Parson. 1992. "Appraising the Earth Summit." *Environment* 34, no. 8 (November): 7–11, 26–33.

Hardin, Garrett. 1968. "The Tragedy of the Commons." *Science* 162 (13 December): 1243–1248.

Hardin, Russell. 1982. *Collective Action.* Baltimore: Johns Hopkins University Press.

Hoel, Michael. 1991. "Global Environmental Problems: The Effects of Unilateral Actions Taken by One Country." *Journal of Environmental Economics and Management* 20: 55–70.

Houghton, J. T., G. J. Jenkins, and J. J. Ephraums, eds. 1991. *Climate Change: The IPCC Scientific Assessment.* New York: Cambridge University Press.

Houghton, J. T., B. A. Callander, and S. K. Varney, eds. 1992. *Climate Change 1992: The Supplementary Report to the IPCC Scientific Assessment.* New York: Cambridge University Press.

Jorgenson, Dale W., and Peter J. Wilcoxen. 1991. "Reducing U.S. Carbon Dioxide Emissions: The Cost of Different Goals." Discussion paper 1575, Harvard Institute of Economic Research, Harvard University, October.

Joskow, Paul. 1991. "Implementing the Tradeable Allowance System for Acid Rain Control." Seminar presentation to Project 88 Seminar, John F. Kennedy School of Government, Harvard University, 2 October.

Kreps, David M. 1991. *Game Theory and Economic Modeling.* Oxford: Oxford University Press.

Lister, Frederick K. 1984. *Decision-Making Strategies in International Organizations: The IMF Model.* Vol. 20, no. 4. Denver: Monograph Series in International Affairs, Graduate School of International Studies, University of Denver.

MacNeill, Jim, Peter Winsemius, and Taizo Yakushiji. 1991. *Beyond Interdependence: The Meshing of the World's Economy and the Earth's Ecology.* New York: Oxford University Press.

Maler, Karl-Goran. 1991. "The Acid-Rain Game 2." Stockholm School of Economics, June. Mimeo.

Manne, Alan S., and Richard G. Richels. 1990. "CO_2 Emissions Limits: An Economic Analysis for the USA." *The Energy Journal* 11, no. 2: 51–85.

———. 1991. "Global CO_2 Emission Reductions: The Impacts of Rising Costs." *The Energy Journal* 12, no. 1: 87–107.

McDonald, Alan. 1990. "Striking Global Environmental Bargains." Paper presented at "North–South, East–West: Establishing a Common Agenda" conference at Institute for International Studies, Brown University, 18–19 April.

Messick, David M., and Keith S. Cook. 1983. "Fairness, Preference, and Fairness Bias." In *Equity Theory*, eds. D. M. Messick and K. S. Cook. New York: Praeger.

Mitchell, Ronald. 1993. "Intentional Oil Pollution of the Oceans." In *Institutions for the Earth: Sources of Effective International Environmental Protection*, eds. Haas, Keohane, and Levy. Cambridge: MIT Press.

Moomaw, William, and Allen Hammond. 1990. "A Modest Proposal to Encourage Unilateral Reductions in Greenhouse Gases." Tufts University. Mimeo.

Nitze, William A. 1990. *The Greenhouse Effect: Formulating a Convention*. London: Royal Institute of International Affairs.

Nordhaus, William D. 1990. "To Slow or Not to Slow: The Economics of the Greenhouse Effect." Cowles Foundation discussion paper, Economics Department, Yale University.

Olson, Mancur. 1965. *The Logic of Collective Action*. Cambridge: Harvard University Press.

Olson, Mancur, and Richard Zeckhauser. 1966. "An Economic Theory of Alliances." *Review of Economics and Statistics* 48, no. 3: 266–279.

Ostrom, Elinor. 1990. *Governing the Commons*. New York: Cambridge University Press.

Parson, Edward A. 1993. "Protecting the Ozone Layer: The Evolution and Impact of International Institutions." In *Institutions for the Earth: Sources of Effective International Environmental Protection*, eds. P. M. Haas, R. O. Keohane, and M. A. Levy. Cambridge: MIT Press.

Parson, Edward A., and Richard Zeckhauser. 1993. "Cooperation in the Unbalanced Commons." In *Barriers to the Negotiated Resolution of Conflict*, eds. K. Arrow, R. Mnookin, L. Ross, A. Tuersky, and R. Wilson. New York: Norton.

Perry, Martin. 1982. "Oligopoly and Consistent Conjectural Variations." *Bell Journal of Economics* 13 (spring): 197–205.

Sand, Peter H. 1990. *Lessons Learned in Global Environmental Governance*. Washington, DC: World Resources Institute.

Schelling, Thomas C. 1960. *The Strategy of Conflict*. Cambridge: Harvard University Press.

———. 1978. *Micromotives and Macrobehavior*. New York: Norton.

———. 1992. "Some Economics of Global Warming." *American Economic Review* 82, no. 1 (March): 1–14.

Sebenius, James K. 1991. "Designing Negotiations toward a New Regime: The Case of Global Warming." *International Security* 15, no. 4: 110–148.

Starrett, David A. 1972. "Fundamental Nonconvexities in the Theory of Externalities." *Journal of Economic Theory* 4: 180–199.

Starrett, David A., and R. Zeckhauser. 1974. "Treating External Diseconomies: Markets or Taxes?" In *Statistical and Mathematical Aspects of Pollution Problems*, ed. John W. Pratt. New York: Dekker.

Subak, Susan, and W. C. Clark. 1990. "Accounts for Greenhouse Gases: Toward the Design of Fair Assessments." In *Usable Knowledge for Managing Global Climatic Change*, ed. W. C. Clark. Stockholm: Stockholm Environment Institute.

Sugden, Robert. 1986. *The Economics of Rights, Cooperation, and Welfare*. Oxford: Blackwell.

Taylor, Michael D., and Hugh Ward. 1982. "Chickens, Whales, and Lumpy Goods." *Political Studies* 30, no. 3: 350–370.

UN Environment Programme. 1991. *Register of International Treaties and Other Agreements in the Field of the Environment*. Report published by UNEP/GC.16/Inf.4. Nairobi: UNEP, May.

UN General Assembly. Intergovernmental Negotiating Committee for a Framework Convention on Climate Change, 1992. "Framework Convention on Climate Change," UN doc. A/AC.237/18. Part II/Add.1, 15 May.

U.S. Central Intelligence Agency (CIA). 1992. *The World Factbook 1992*. Springfield, VA: National Technical Information Service.

U.S. Task Force on a Comprehensive Approach to Climate Change. 1991. "A Comprehensive Approach to Addressing Potential Climate Change." Presentation to the UN INC meeting, Washington, DC, February.

Victor, David G. 1991. "How to Slow Global Warming." *Nature* 349 (7 February): 451–456.

Weitzman, Martin L. 1974. "Free Access vs. Private Ownership as Alternative Systems for Managing Common Property." *Journal of Economic Theory* 8: 225–234.

Wirth and Heinz, 1991. "Project 88—Round 2. Incentives for Action: Designing Market-Based Environmental Strategies." Public policy study sponsored by Senators Timothy E. Wirth and John Heinz, 1991.

World Bank. 1992. *World Development Report, 1992.* New York: Oxford University Press.

World Resources Institute. 1990. *World Resources, 1990–91.* New York: Oxford University Press.

Young, H. P. 1990. "Sharing the Burden of Global Warming." School of Public Affairs, University of Maryland, November. Mimeo.

4 Improving Compliance with the Climate Change Treaty

■ **RONALD B. MITCHELL**
Department of Political Science
University of Oregon

■ **ABRAM CHAYES**
Harvard Law School
Harvard University

Most of the analytic and negotiating energy surrounding the development of a climate change treaty has focused on substantive limitations on net emissions of greenhouse gases, whether through targets and timetables, emissions permits, taxes, or technological standards. But no matter how stringent these commitments are, the treaty will not succeed unless the parties comply with them. The compliance problem must be addressed from the outset and a compliance system must be designed into the treaty from the beginning. In the climate change arena, the costs and magnitude of required behavioral changes, and the regulatory breadth and complexity, pose especially difficult compliance problems. Moreover, unlike most international agreements, which seek to affect only the actions of governments, a successful climate change treaty must alter the behavior of ordinary individuals and business firms whose activities account for the emission of greenhouse gases. The treaty must encourage national governments not only to comply with its provisions by adopting legislation and other appropriate policies, but also to take action to facilitate compliance and condemn violations of private actors within their own borders.

Numerous international and national factors influence whether a nation complies with a given rule of a given treaty at a given time—for example, the distribution of international power or the administrative efficiency and constitutional structure of the parties.[1] These factors bound the aggregate level of compliance that can be expected under a treaty. This chapter focuses on the international rules, policies, and processes through which international organizations, states, corporate entities, and nongovernmental actors can influence the level of compliance with a specified treaty.

With respect to compliance, nations and subnational actors can generally be divided into three categories. First, some nations will have incentives and abilities that lead them to comply with a given treaty rule independent of the systems established to identify and respond to noncompliance. Thus, for example, many OECD countries have already committed themselves to reducing their greenhouse gas emissions. Given their domestic environmental concern and their ability to pay the costs involved, these states seem likely to reduce greenhouse gases on the schedules and in the ways required by international regulation.

Second, for some nations compliance will be contingent on the type and likelihood of responses to noncompliance institutionalized in the international regime. Included in these nations are those that desire to comply but lack the financial and administrative resources to do so and those that believe themselves to be marginally better off by not complying with the treaty so long as no significant costs attend violation and no major financing is provided for compliance.

Third, some nations will have incentives and abilities that will lead them either to not sign an agreement or to sign and violate the agreement independent of the systems established to identify and respond to noncompliance. Malaysia's opposition to certain terms of the forestry principles that were under negotiation at UNCED in June 1992 is one example.[2]

Which of these categories a state falls into is not foreordained. Where a nation falls depends on how the rules are framed, the direct costs and benefits of compliance, the expected actions of other international actors, domestic political forces, the degree of dependence on other states, and the nation's infrastructure and resources. Whether actors in the second category actually comply will depend on the structure of the compliance information and response systems. Thus, effective implementation requires a coherent and integrated approach that addresses all three types of actors. An integrated approach will make initial compliance likely, will identify and respond appropriately to noncompliance when it occurs, and will recognize those actors whose behavior will remain unchanged under the treaty and make the other elements of the system robust to such free-riding.[3]

This chapter examines the policy options available for the creation of a successful compliance system for a climate change regime and the obstacles that stand in the way. We organize our analysis around the three questions suggested above.

- How can negotiators design the rules and procedures to elicit the highest possible compliance levels?

- What kind of information and monitoring system will provide information about the status of treaty-related behaviors?

- What kinds of response are appropriate in cases in which the information discloses a failure to comply with treaty commitments?[4]

We then discuss the institutional structure needed to implement an effective compliance system and assess the likelihood that the FCCC signed in Rio de Janeiro in 1992 will foster the development of such a structure. We conclude by recommending that policymakers design, and dedicate resources to, a compliance system that emphasizes efforts to make compliance easier for actors predisposed to comply while seizing less frequent opportunities to alter the incentives and capacities of those disinclined to comply.

RULES TO MAXIMIZE THE PREDISPOSITION TO COMPLY

Most discussions of what greenhouse gases to regulate are based on the contributions of a gas to global warming.[5] They are also based on which economic sectors produce the greatest share of a given gas.[6] To the extent that these discussions address compliance at all, they usually specify a gas or activity to regulate independent of compliance considerations and assume that monitoring and enforcement systems can be developed to maximize compliance. But the effectiveness of a rule in regulating climate change depends not only on its potential for reducing greenhouse gases under conditions of perfect compliance, but also on the level of actual compliance.

The choice of what gases to limit, which sources of emissions to regulate, and how to frame the regulations determines who must comply and how much compliance is likely. That choice, in turn, requires answers to two subsidiary questions: First, which economic actors and sectors responsible for emissions of a given greenhouse gas are most likely to alter their behavior in response to national efforts to legislate, implement, and enforce international treaty commitments? Second, which nations are most likely to have the incentives and ability to legislate, implement, and enforce such policies?

FIGURING WHICH SECTORS ARE LIKELY TO COMPLY

The most obvious determinant of an actor's compliance is the cost of adjusting behavior to meet the mandated requirements: the net cost of the changes required to come into conformance with the rule, the cost of alternative means of achieving the economic goal at lower emissions levels, and the value of any benefits from the shift in behavior. Companies or individuals with large resources will be more capable of bearing such costs, and, if they are high-visibility actors concerned about their public reputations, as with many multinational corporations, they will also have stronger incentives to do so. Producers can ordinarily pass on the costs to consumers and thus have fewer disincentives to comply. Actors facing less economic competition, and those who can readily perceive violations by competitors, will be more likely to comply because their own compliance is less likely to place them at an economic disadvantage. Finally, actors who are already regulated in other arenas or involved in existing informational infrastructures may be more likely to comply because they are more likely to be aware of the rule and its requirements and to be embedded in a culture and habit of compliance. Public utilities in the United States, for example, already face considerable regulation, including extensive monitoring, and stay abreast of new regulations as they are promulgated. In contrast, farmers in many developing countries may face very few regulations and be unaware of those they do face because they lack access to newspapers and other means of informing them of regulations, which are often taken for granted in other settings.

In regulating emissions of a given gas, policymakers may have wide latitude in selecting a point in the pollution production process, and therefore which actors, to regulate. As one example, electricity-related carbon dioxide emissions could be reduced by limiting use of fossil fuels by power companies, by mandating efficiency standards for appliance manufacturers and building contractors, or by requiring power consumers to conserve energy. Actors farther upstream in the pollution production process generally have fewer disincentives to compliance.[7] Compliance would be facilitated most by limiting the fossil fuel use of power companies, which most countries strictly regulate and which are quasi-monopolies that can readily pass on increased costs to consumers.

INDUCING NATIONS TO IMPLEMENT TREATIES

A country's compliance decision will be driven in part by the perceived costs and benefits of implementing and enforcing their international obligations. Climate change commitments may provide the additional impetus for some states to undertake actions that have other significant benefits but that were not otherwise politically possible. Thus, for example, a climate treaty may make gasoline taxes, which can reduce energy dependence and provide revenues, more attractive in energy-importing states. A country's ability to pay the costs of compliance also will influence the compliance decision, so that developed countries may be more likely than developing countries to meet their obligations. Domestic politics will shape the perception of whether the benefits of a regulation outweigh its costs, so that the breadth and power of domestic environmental lobbies relative to those actors targeted by the regulation will be an important factor.

Other considerations will determine whether a regulation, once adopted, is likely to lead to national compliance with international commitments. Different regulatory approaches may have quite different outcomes. For example, even though actors may be more likely to comply with a carbon tax than with command-and-control or emissions-trading policies, taxes are inherently less certain with respect to total emissions and may therefore lead to inadvertent national-level noncompliance.[8] Second, different political cultures may make what is apparently a more effective regulatory system politically unacceptable in a country. Thus, while Finland and Norway have adopted carbon taxes, the United States has found it difficult to adopt such a policy.[9] Third, many countries, especially in the developing world, may simply lack the administrative and informational infrastructures necessary to translate well-designed legislation into real behavioral change. Akin to the efforts for transferring "appropriate technology" during the 1970s, the 1990s may see a need for "appropriate environmental policies" that take into account the informational and resource limitations that inhibit regulatory efforts in developing countries. The power to control relevant economic actors, which is often taken for granted in industrialized countries, may be doubtful when small developing nations seek to control the actions of multinational corporations.

Successful regulations must take into account compliance factors at the actor, sector, and national levels. For example, recent policies and negotiating positions suggest that the interests and capacities of the members of the OECD make them significantly more likely to comply with climate change treaty commitments than developing countries and former members of the Soviet Union. If this inference is true, a treaty requiring reductions in a greenhouse gas primarily produced by OECD countries (for example, carbon dioxide) will produce greater actual greenhouse gas reductions than would an otherwise equivalent treaty that required reductions in a greenhouse gas primarily produced by developing states (for example, methane).

For example, methane emissions from rice cultivation occur mainly in China and India, two countries unlikely to comply with climate change rules without large financial transfers. However, the regulation of oil consumption will probably elicit higher compliance: oil companies are more regulated and better informed of new regulations, and they operate in countries more likely and able to enforce fuel consumption targets or carbon taxes.

The point here is not to recommend a specific regulatory form, but rather to encourage careful comparison of alternatives on the basis of their compliance implications, avoiding the frequent assumption that the level of compliance will be equal for all possibilities. Compliance is rarely the only, or even the primary, goal of policymakers. Indeed, they will assuredly promulgate limits on certain activities despite low compliance. The framework presented here can help provide policymakers with a general assessment of the likelihood of compliance for alternative policies. This assessment can help them select activities for initial regulation, so that the climate change regime will develop a record of successful regulation and thus foster subsequent regulatory efforts.

AN EFFECTIVE COMPLIANCE INFORMATION SYSTEM

An effective compliance information system (CIS) facilitates compliance with a climate change agreement in several ways. First, it helps distinguish noncompliance due to inadvertence and lack of attention to the issue from willful violation. Second, the threat of peer pressure and adverse world public opinion implicit in the collection and dissemination of information may act as a deterrent to violation. Indeed, the Sustainable Development Commission relies exclusively on publicizing noncompliance data to foster implementation of Agenda 21.[10] Third, an effective CIS identifies those actors uninfluenced by these first two factors, so that measures can be taken to bring them back into compliance.

To achieve these objectives, the CIS must have four components: rules that can readily be verified, a national-level self-reporting system, some form of independent verification system, and a capability to analyze the data that are generated.[11]

DEVISING RULES AMENABLE TO VERIFICATION

The transparency or verification suitability of a given activity that contributes to global warming depends on the nature of the activity and the ability to observe that activity.[12] Factors that influence ease of verification include the number of sources of a given gas; their size, mobility, and accessibility; the concentration of sources; the continuity of emissions; the ease and accuracy with which emissions may be imputed from factor inputs; and the technical capacity to measure emissions.[13] Ideally, regulated emissions would be confined to a few large, concentrated, stationary, readily observable sources that produce continuous emissions which vary little for a fixed amount of inputs. This model suggests not only which gases to regulate to achieve transparency but also where in the emissions cycle to regulate.[14] For example, although the Montreal Protocol nominally limits both production and consumption of ozone-depleting substances, the secretariat will collect data only on national production of, and trade in, CFCs because these are far more transparent than CFC consumption.[15]

The form of the treaty norm will also influence transparency. As arms control negotiators have argued for years, banning an activity completely rather than specifying a numerical limit makes the detection of noncompliance much easier.[16] Infractions of quantitative limits, like those in the Montreal Protocol, are more observable than qualitative limits, like the Wetland Convention's requirement that signatories make "wise use" of wetlands.[17] Compliance with the FCCC's various requirements that nations "cooperate" to mitigate climate change will prove impossible to verify in the absence of flagrant bad faith. Drawing clear and specific lines between compliance and noncompliance makes the activity not only more transparent to others but also to the actor itself, thus making compliance easier.[18]

Detecting a violation, however, may not be equivalent to identifying a violator. Negotiators often paper over conflicts by using the passive voice to avoid assigning responsibility for meeting an obligation to a particular actor.[19] A climate change treaty may bind national governments to meet emissions targets, but some governments may have only limited direct control over the largely corporate and individual actors responsible for those emissions.[20] Rules that link fossil fuel consumption with emissions may prove sufficiently accurate to identify national-level breaches of emissions targets, but they are unlikely to prove adequate in most legal systems to authorize prosecution of an individual power plant.[21] This situation hinders efforts to identify noncompliance. While treaties generally place obligations on governments, these obligations often must be performed by private corporate and individual actors. Treaties could facilitate compliance by prescribing or proscribing specific actions—for example, a ban on new constructions of coal-fired power plants. While unlikely at present, such regulations would permit easier identification of violations and violators.

ESTABLISHING A SELF-REPORTING SYSTEM

Self-reporting is the primary basis for most existing environmental treaties. It is a major component of the FCCC and the Sustainable Development Commission.[22]

A successful self-reporting system would elicit from all member nations prompt and regular reports with high-quality, compliance-relevant information that is comparable over time as well as across nations. The system should seek to develop a time-series database that identifies baseline levels for activities that already are, or might reasonably become, regulated. Since for many gases compliance will be defined relative to a specified baseline, the data needed to identify that baseline must be developed. To this end, the reporting of historical data should be encouraged. Data collection procedures should be standardized to facilitate reporting and to achieve comparability.[23] Reports should reflect government enforcement against private actors for violations of domestic legislation implementing the treaty. Reports should identify the sources of noncompliance so that proper policy responses can be developed.

The need for a high-quality reporting system has become part of the conventional wisdom.[24] However, several recent studies have demonstrated that compliance with the reporting provisions of environmental treaties cannot be assumed.[25] Little attention has been paid to how the amount and quality of international reporting can be maximized. Indeed, reporting provisions pose compliance problems that require increasing the incentives and ability of nations to report.

INCREASING THE INCENTIVES AND CAPACITY TO REPORT

A self-reporting system must overcome inherent disincentives of the members to report their own violations or noncompliance accurately. To this end, the treaty organization must seek to demonstrate the importance of frequent, high-quality reporting to the goals of the treaty.[26] Ensuring that discussion and analysis of the reports occur at annual meetings can help, as can high-quality use of the data in annual reports. The reporting format should require governments to explain the reasons for any failures to meet treaty norms. States are more likely to report noncompliance when they believe that it will be met with efforts to assist in their achieving future compliance rather than with sanctions for past noncompliance.

Achieving a high-quality reporting system requires that a standardized report format be developed early and thoughtfully. Questions should be framed in terms that allow subsequent verification of compliance behavior. All necessary data should be requested explicitly.[27] The report format should allow comparison of data across countries. In a climate change agreement, this provision for comparability would require reporting on the sectors covered in emissions estimates and the basis for measurement and conversion.[28] The data to be collected and the conversion factors to be used should be explicitly delineated on the form itself.[29] States should be required to report even during periods of nil activity so that noncompliance with reporting provisions can be distinguished from substantive noncompliance.[30]

To the extent possible, the reporting system should be integrated into the standard operating procedures of the responsible national bureaucracies and should rely on modifying existing domestic information infrastructures rather than on creating new ones. Combining a requirement that nations designate offices for reporting data with an interactive computerized reporting system would greatly improve both the quality and the amount of reporting.[31] Computers and modems provide a direct link and a sense of a real presence of the international organization within national governments, raising the perceived importance of reporting and the dedication to it. Developed countries could provide developing countries with needed hardware and technical assistance. A computerized system would reduce the need to disseminate forms and, if coupled with an electronic mail system, would allow interactive questioning to correct obvious errors, clarify ambiguities, and secure additional information. An electronic system would also make it easier to improve the clarity and usefulness of the form.[32] Finally, reporting in electronic formats would encourage compliance by making analysis and dissemination easier, and hence more likely.

Regardless of how well the information system is designed, major problems will exist in the ability of nations to compile the necessary data on emissions. For example, many obstacles exist to the estimation of emissions of gases from large-area sources, such as biomass burning or methane from rice cultivation, and accurate estimation is almost impossible.[33] The numerous complex data requirements of a climate change treaty make it likely that many countries will not be able to provide many data elements for some time. Countries should be encouraged to identify those data elements that they are unable to provide and the obstacles they encountered.

States frequently fail to report because of a lack of resources or a lack of attention to the issue. Developed countries should provide technical advice so that the information-intensive requirements of a climate change agreement can be met in developing countries that currently lack the necessary information infrastructure and resources. The Montreal Protocol provides for technical and financial assistance that could be used to improve required reporting.[34] Claims for financial assistance could also be made conditional on good faith efforts to comply with reporting requirements for some specified number of years prior to receiving funding.

VERIFYING NATIONAL REPORTS

Because the incentives against self-incrimination will lead to some failures to report or to false reports, an independent verification system is a necessary element of an effective CIS. Political and practical constraints dictate that such a system be limited primarily to identifying major treaty breaches. Four different models for the independent identification of noncompliance with environmental treaties either have been proposed or are in use. The systems could be used together.

The first model involves combining "national declaration systems and an international verification system which can largely dispense with measuring systems."[35] The regime on Long-Range Transboundary Air Pollution has successfully used such a system.[36] The LRTAP system compares national reports of sulfur and nitrogen oxide emissions to its organization's own emissions calculations on the basis of analysis of statistics on fossil fuel consumption.[37] This method has been frequently proposed as the major means of verifying compliance with the carbon dioxide component of a climate change agreement. Whether such independent verification can be used will depend on the types of gases and activities regulated by the regime.

A second model for independent identification of noncompliance uses cooperative on-site monitoring procedures when information developed through self-reports, satellite observation, NGOs, and the like raises concern among other parties. The Sustainable Development Commission includes plans for such a system.[38] Parties would agree in advance to an on-site inspection in such cases. If inspections were not perceived as punitive but rather as a way to assist nations to meet their international commitments, acceptance of the request would appear likely. The secretariat for the Wetlands Convention has conducted over twenty voluntary on-site wetlands inspections since the monitoring procedure was established in 1989. No request under this procedure has yet been refused.[39]

A third verification model is the use of nonintrusive data collection that does not require the cooperation of the suspect country.[40] Satellites and reconnaissance aircraft have been frequently proposed as providing means for identifying major breaches of a deforestation treaty or a climate change agreement (for example, as a way to measure emissions due to agricultural activities). While several countries already have remote-sensing satellite capabilities or will soon, many obstacles prevent their use for accurate verification of reported data or for independent detection of extensive clear-cutting or other land-use changes.[41] Numerous legal, economic, and political questions also may hinder an international organization from gaining access to satellite, aerial, or land-based data obtained by national governments or private companies. Satellites may prove most useful in connection with agreed inspection systems used in the second model described above.[42]

Fourth, if international concern about the impacts of global warming becomes sufficiently intense, it may be possible to institute challenge inspections that sacrifice some national sovereignty to achieve greater international environmental protection. Major movement in this direction appears unlikely at present, but if and when opportunities arise, it will be important to take advantage of them to begin to change the international legal system.

Apart from the treaty organization, individual states, through the operation of national statistical or intelligence services, may develop information that can be used to verify reports submitted under treaty requirements. Similarly, NGOs can supplement these systems. NGOs and industries often have the incentives, the financial resources, and the informational infrastructure to identify cases of noncompliance. Information generated by national governments or by NGOs must be checked against less partisan sources, but it provides an important method for priming the informational pump.[43]

ANALYZING COMPLIANCE INFORMATION

None of the data collected, whether from national reports or independent sources, will facilitate compliance unless they are analyzed to determine whether violations occurred, and if so, what the causes were and who was responsible. Such an analysis is often beyond the capabilities of the treaty organization. Significant effort is also needed to make the data comparable across countries and across time, so that trends can be identified and marked deviations noted.[44] Effective analytic capability implies an ongoing process of inquiry to clarify confusions and ambiguities, to raise the salience and importance of reporting and compliance, and to help avoid false accusations of violation.

It is often a matter of controversy whether the data produced show that the country is not in compliance. To give one example, in June 1992 NASA released satellite photos of the U.S. Pacific Northwest, claiming that clear-cutting was threatening "the ability of the forests to support a diversity of species." The USDA Forest Service responded that "it is misleading to make judgments about forest practices based on pictures from space."[45] Thus, even domestically, satellite data may be open to significantly different interpretations. Decisions about whether specific information is sufficient to warrant a response or even further investigation are ordinarily made by the political organs of the treaty organization.

The analytic process itself is not just a matter for the organization's staff. Compliance reports and analyses must become an important agenda item at regular meetings. The integration of this information into meeting structures is essential both to emphasize the importance of reporting and compliance and also to ensure that a regular process is established by which the sources of noncompliance can be identified and corrected. The process should include provision for disseminating the results as widely as possible. Publicity is the key basis on which positive political rewards will be bestowed on those fulfilling their international commitments, on which assistance will be provided to those failing to do so because of a lack of resources, and on which sanctions will be brought to bear on those intentionally violating the terms of an agreement. The Sustainable Development Commission's planned reliance on publicizing noncompliance information confirms the importance placed on dissemination as an essential element in altering behavior.

THE REMAINING OBSTACLES

Several factors are likely to prevent the development of a fully effective CIS, despite the best efforts of the treaty organization. The inherent disincentive to self-incrimination poses a continuing obstacle to a successful self-reporting system. Compliers and noncompliers seeking financial assistance may well report. Intentional noncompliers and nonsignatories will not.

Technical and financial obstacles to conclusive, nonintrusive data collection and political resistance to inroads on sovereignty will likely preclude intrusive inspections necessary to verify compliance. Reliance on rule-of-thumb calculations rather than on accurate emissions measurement, as will be required for some time given current measurement technologies, will also hamper the identification of states that have violated the agreement. Similarly, the definitions of compliance will likely be too vague, at least during the initial years, to support charges of noncompliance.

In short, given the low costs of violations to the regime and the high costs of detecting them, it will not be surprising if the initial years of a climate change treaty are spent in trying to refine the rules rather than to ensure compliance with those already established.

THE NONCOMPLIANCE RESPONSE SYSTEM

Noncompliance due to inadvertence or incapacity and noncompliance caused by more or less deliberate violation of the agreement are two different kinds of conduct and suggest quite different responses. Noncompliance should be met with positive efforts to facilitate and encourage compliance by removing knowledge and resource barriers, by reducing the costs of compliance, or by increasing the ability to pay them. Violation might warrant some form of sanctions to alter the incentives and make violation less attractive.[46]

These two types of response can also have quite different actors as their targets. The compliance system has both a national and a subnational component. Responses either can seek to get nations to take steps to induce compliance by their own corporate and individual citizens or, in certain cases, can seek to directly affect the choices of nationals of other countries through market-based measures, including establishing direct links between compliance and market access.[47]

FACILITATING AND ENCOURAGING COMPLIANCE

Inadvertence (mistakes, lack of knowledge of requirements, or inadequate capacity to fulfill them) can be a major source of noncompliance. The response to this type of noncompliance must address its source.

Actors, be they nations, corporations, or individuals, are often unaware of the existence, nature, and cheapest means of complying with new regulations, especially when the regulated activities involve numerous, widely dispersed actors unconnected to existing information infrastructures. Thus, one source of noncompliance with a deforestation treaty would certainly be the failure to inform the poor and often illiterate peasants who cut trees for firewood and who clear agricultural land. Even if they are aware of new domestic limits to tree cutting, compliance will be unlikely unless they can be educated about culturally consistent alternatives for making a living without cutting trees. Monitoring procedures involving site visits, training programs, and seminars may provide opportunities for informing the responsible actors of what their commitments are and how they can most readily comply with them.

Valuable forums for communication can be provided by conferences of bureaucrats responsible for national implementation of a treaty; representatives from economic sectors that must comply; legal, policy, and technical experts; and treaty organization staff. Interchange between those responsible for devising rules and procedures to regulate climate change and those responsible for enforcing and complying with them is an important first step toward removing the barriers to compliance.

Compliance can be facilitated by the provision of technical advice to countries that are making efforts but that are nonetheless failing to achieve compliance.[48] As with the monitoring procedure, technical advisers with broad field experience can help countries devise cheaper and more effective means of achieving compliance and of conducting national-level enforcement activities. Lessons learned in one country may be applicable in others. Similarly, when new technologies become available, at-cost transfers to developing countries can ensure that they are adopted as widely as possible.

The most frequently discussed and most expensive means of encouraging states to comply is direct financial transfers conditioned on nations undertaking activities that will improve compliance. A modest compliance fund has been established under the Montreal Protocol. Funding might be provided for the technology transfers discussed above, for administrative programs to improve the effectiveness of human resources, or for capital projects. One frequent suggestion is that an international fund be established to which industrialized states would contribute and from which developing states would receive funds to cover the costs of compliance. Financing could also occur bilaterally, perhaps with an international organization as a clearinghouse, although this would probably mean fewer funds for developing countries. In any case, the magnitude of the required funding will be a major political issue in the legislatures of developed countries.

Thus far, developed countries have been unwilling to make commitments to provide funding to ensure compliance by developing states.[49] Even if they did, these commitments would face the same type of compliance and enforcement problems discussed in this chapter. The Sustainable Development Commission will address this problem by monitoring compliance with the financial and technology transfer provisions as well as with the substantive provisions of Agenda 21 and other environmental agreements.[50]

Businesses may prove a valuable source of resource and technology transfers to developing countries. If national and international laws create positive incentives for businesses to finance specific projects that reduce greenhouse gases or to provide at-cost or free technology and capital equipment, they could provide a valuable source of funding. Regulations that require businesses to reduce their own emissions or finance an equivalent emissions-reducing project in another country may prove politically acceptable where direct funding by the taxpayer would not.

Finally, NGOs have shown themselves willing and able to organize financial resources for transfer to developing countries in exchange for more environmentally beneficial programs, for example, debt-for-nature swaps.[51] International treaties could facilitate such transfers by incorporating provisions that create a clearinghouse for requests for and offers of assistance and by removing obstacles that make banks reluctant to agree to the terms of such agreements.

In any case, those who fund projects to help actors comply need assurances that the project's terms will be met. Whether funded by an international climate change fund, the World Bank, a corporate tree-planting project as part of an emissions-trading scheme, or an NGO debt-for-nature swap, the funders will want to know whether the project was properly implemented. Corruption and inefficiency in some countries and the difficulty of auditing the use of funds make it very likely that some projects will not be completed, will run over budget, or will not result in the desired improvement in compliance. In short, efforts to facilitate and encourage compliance must be sufficient in size and appropriate to the source of the noncompliance. They must also lead to the expected changes in behavior.

SANCTIONING VIOLATIONS

Many authors bemoan the absence of strong enforcement provisions in a climate change treaty as well as in other environmental accords. They want a treaty with teeth.[52] If "sanctions" refers only to coercive economic and military measures to punish past transgressions and deter future violations, in the manner of domestic criminal law, the likelihood of a treaty with teeth is slight.[53] But sanctions may denote a wider array of possible responses. They can be either formal or informal, can be imposed by both state and nonstate actors, and can consist of nothing more than having to respond to questions regarding the alleged violation.

Increasing the perceived costs of violations begins with publicizing the violations when they occur. One of the most common forms of jawboning and shaming is the discussion of infractions at annual meetings of parties to the treaty, which requires the country concerned to explain and justify its conduct. The Sustainable Development Commission appears to be focused on just such efforts.[54] Publicity coupled with strong domestic political pressures to appear "green" led many states to adopt more environmental positions after ministerial meetings under the North Sea agreement.[55] The blacklisting of violators can prove costly to corporations that value their public image or states that want good relationships with their neighbors.[56] At the subnational level, NGOs often disseminate information on violations through their publications and the news media, which can lead to consumer boycotts of an offending country or company.[57]

Despite the obvious sanctioning opportunities that publicity provides (or perhaps because of them), many international environmental organizations do not publicize information regarding countries or companies that fail to meet their treaty commitments. For example, the parties to the Memorandum of Understanding on Port State Control have consciously chosen not to make public the names of oil tankers caught violating oil pollution treaty regulations because of concerns over liability for false information and potential antitrust implications. They also conspicuously avoid publishing state-by-state inspection statistics in annual reports, presumably because of diplomatic unwillingness to embarrass other countries, despite the Memorandum of Understanding requirement that states inspect 25 percent of the ships entering their ports.[58] On the other hand, it is equally important to verify data provided by NGOs before using them as the basis for negative publicity. Thus, although publicizing data on noncompliance may be an important form of deterrence, legal concerns and political barriers to the use of such data must be addressed.

Selective taxes provide another means to increase the costs of violations. Taxes at the domestic level can serve as "graduated" penalties to deter production of a given greenhouse gas by increasing the costs of producing the emissions. Carbon taxes, which have been heralded as a means of inducing private actors to decrease their greenhouse gas emissions, seem more likely to be used as a domestic rather than an international policy tool. A frequently touted advantage of a carbon tax is that it can piggyback on existing domestic tax collection systems, interceding at points of existing market transactions.[59] At the international level, however, no such infrastructure exists, and it seems highly unlikely that states will grant such taxation powers to an international body. Similarly, some authors have proposed that fines should be used to enforce emissions limits.[60] In the domestic context these may be appropriate, although often they are not the most effective way of inducing compliance with regulatory requirements. At the international level, however, there is no adjudicative or collection infrastructure to implement such a measure and little likelihood that the power to impose fines would be granted to an international body.

One way of dealing with these difficulties would be a treaty provision that a multinational corporation could be taxed in its "home" country for activities conducted elsewhere if the host country failed to tax it for that activity. Similar arrangements could be instituted for fines for violations. Such provisions would increase incentives for host countries in the developing world to tax or fine companies since, if they did not, the developed home country would. This taxation method would also allay the fears of home countries that levying such taxes or fines would place a host country at a competitive advantage. These measures would be available, however, only between states that were parties to the treaty.

Environmental treaties often impose limitations on trade in goods that are prohibited by the treaty, such as endangered species or CFCs. At least nineteen environmental treaties authorize some form of trade measures in support of their provisions.[61] Taxes could be imposed on imports from countries failing to sign or comply with a climate change agreement.[62] In the case of major multinational corporations found in violation of an agreement, similar taxes could be imposed on their imports. Such sanctions would raise serious problems, however, not only under the GATT, but also under bilateral commercial treaties and other agreements providing for most-favored nation status or national treatment of imports. While various proposals were made for the climate change convention to provide for trade sanctions against nonsignatories and noncompliant parties, it is not surprising that the final FCCC did not include such provisions. Widespread use of trade sanctions to enforce environmental agreements is unlikely until nations resolve the larger issue of free trade versus environmental protection.

It has been suggested that the World Bank and the IMF might discontinue loans or progress payments on the basis of compliance with treaty rules on climate change. The World Bank's rules already provide that it will not finance a project that conflicts with an international environmental obligation of the borrowing state. However, broader financial sanctions would probably require changes in the charters of the lending institutions, which now require decisions to be based solely on economic considerations.

In any case, experience with economic sanctions in other issue areas shows that they do not provide an easy mode of enforcing international obligations.[63] Ordinarily, the ability to impose economic costs large enough to be effective depends on widespread cooperation among a large number of states and on the inability of the sanctioned state to skirt these costs. It is very difficult to mobilize this kind of cooperation, and even then, as the case of Iraq shows, the target country can hold out for a long time.

The most effective sanctions are likely to be those that are directed at violating companies, rather than at countries, and that prevent the company from conducting business in a country that is one of its major markets. Fines and taxes can be incorporated into the cost of doing business, but the limitation of operational access to major markets poses a far more potent threat. For example, many tanker operators continued to discharge oil at sea even though this practice subjected them to the possibility of a fine, but the threat of having their tankers detained in port or denied entry for noncompliance led most tanker owners to comply with expensive requirements for segregated ballast tanks and other equipment.[64] Sanctions imposed directly against businesses skirt contentious sovereignty issues, which often provide the rationale, if not the reason, for states to avoid sanctioning violations.

Many states will be reluctant to undertake sanctions either because they are viewed as ineffective or because the costs of imposing the sanctions appear too great. Some countries, on the other hand, have shown a strong willingness, usually because of strong domestic pressure from environmentalists, to sanction those actors that threaten the effectiveness of an environmental treaty. U.S. threats of sanctions against nonmember whaling states and against states impeding the effectiveness of the Convention on International Trade in Endangered Species (CITES) provide some examples.[65] Many states recently banned wildlife purchases from Thailand because of its violations of the CITES treaty.[66]

Sometimes, states willing to impose sanctions will not do so because of existing legal barriers to sanctions. One example, albeit involving domestic rather than international rules, is the GATT condemnation of a U.S. ban on imports of Mexican tuna because of the high number of dolphins killed. Reducing or eliminating such barriers may be the best way to impose costs on violators. The explicit requirement of the International Convention for Prevention of Pollution from Ships that port states detain ships detected in violation of the equipment requirements did not lead states uninterested in enforcement to detain noncomplying ships, but it did result in detentions by states inclined to enforce the rules by removing the legal barriers that had previously prevented them.[67] We conclude that international treaties can best improve the prospects for sanctions by removing the practical and legal barriers that inhibit states already predisposed to undertake sanctions rather than by attempting to induce reluctant states to do so. New legal rights may prove more effective than new legal obligations.

As with disseminating information, NGOs may feel fewer constraints on imposing sanctions on companies or states found in violation of treaty rules. NGO efforts in the United States have brought about changes in tuna-fishing methods to reduce dolphin by-catch even in the absence of international regulations on the subject. Evidence of violations developed or disseminated by NGOs can provide the basis for effective mobilization of consumer boycotts and other actions.

Finally, business or industry practice can sometimes provide a channel for increasing the costs of violating an agreement. Certifications or insurance premiums linked to historical violation rates or to the conformance of technology to international standards may provide strong incentives for polluting businesses to pay greater attention to international regulations. Similarly, large and powerful green corporations may be capable of using the threat of taking their business elsewhere to make companies with which they do business more responsive to environmental rules.

THE INSTITUTIONAL FRAMEWORK

The compliance information and response system described above involves a number of complex organizational activities.

- Collection and analysis of party reports on performance

- Follow-up inquiries or independent verification activity

- Investigation of the causes of any compliance problems, usually by means of prolonged interactive exchanges with officials in the country concerned

- Development of recommendations for corrective measures, often including identifying sources for necessary funding

- Preparation of reports to the governing body having jurisdiction, together with drafts of appropriate action documents

- Follow-up to ensure that the country concerned has taken the necessary steps to bring itself into compliance

All of the elements above imply a significant level of organizational and bureaucratic resources and competence.

Unfortunately, current experience with international organizations does not bode particularly well for the fulfillment of these organizational requirements. Over recent decades, the international community, and particularly the developed countries that provide most of the resources, has manifested a pronounced skepticism of international bureaucracies and organizations. The payor countries have conducted a continuous campaign to reduce the budgets and activities of existing organizations and to resist the creation of new ones. This resistance has manifested itself particularly in the area of environmental affairs; as a result, the sharp increase in environmental law making over the past two decades has not been accompanied by the development of any overarching international secretariat with broad responsibility over the range of international environmental concerns. The UNEP, created after the UN Conference on the Human Environment in 1972, is a creature of the General Assembly, rather than a specialized agency of the UN, and therefore does not have its own charter and budget. It operates on extremely limited funding and is confined to catalytic and broad review functions.

Most environmental treaties have created their own highly decentralized governing institutions. Secretariats are rudimentary, and reliance is placed on various committees and working groups staffed by the parties themselves for analysis of relevant data and for preparation and documentation of the substantive actions of the organization. The situation is further confused and diffused because of the framework convention and protocol format assumed by many modern environmental treaties, including the FCCC adopted at Rio de Janeiro. Since the parties to any single protocol are not necessarily identical to those of the framework convention itself or of other protocols, each protocol to some extent must have its own governing institutions, although ordinarily the same overburdened, understaffed, and underfunded secretariat acts for all protocols under a single convention.

Thus, for example, the ozone regime operating under the Montreal Protocol has its own Conference of the Parties, separate from that of the Vienna Convention, its parent convention. When, a year after the adoption of the Montreal Protocol, it became apparent that many of the parties were not providing the required reports on imports, exports, and production of controlled chemicals, the Conference of the Parties established an Ad Hoc Group of Experts on Reporting to investigate the causes and to make recommendations for improving the situation. The Conference of the Parties has proliferated similar committees and expert groups, generally staffed by representatives of the parties, to address similarly narrow issues. The secretariat, which must also service the Vienna Convention, consists of less than a dozen people employed by the UNEP in Nairobi, Kenya. By contrast, the Organization for the Prevention of Chemical Warfare, established by the Chemical Warfare Convention, is expected to have a secretariat of 500 or more and an annual budget in the range of $100 million.

It is too early to be sure, but the FCCC, which entered into force in March 1994, seems likely to adopt the organizational approach of the environmental treaties. It establishes a Conference of the Parties, in which all significant powers of the organization are vested. There is a Subsidiary Body for Scientific and Technological Advice and a Subsidiary Body for Implementation. Each is a committee of the whole in the sense that it is made up of governmental representatives rather than independent experts, and any party that wishes can be a member. A financial mechanism to raise and distribute moneys for assisting developing countries to reduce emissions is to be established at the first meeting of the Conference of the Parties. In the interim these functions are to be performed by the Global Environmental Facility administered by the World Bank, with the assistance of the UNEP and the UN Development Programme. It seems likely that this arrangement will be confirmed when the Conference of the Parties meets.

One short article in the FCCC establishes a secretariat with strictly limited functions; in particular, it is to (1) make arrangements for meetings of the policy-making bodies established by the convention, (2) compile and transmit (but not analyze) the reports from the parties, and (3) facilitate (but not provide or arrange for) assistance to parties in developing countries in preparing the reports.

Although practice may in time introduce some latitude of interpretation, the text of the convention is relatively meager. The convention establishes no quantitative targets or limits on emissions, although developed countries "recognize" that "the return by the end of the present decade to earlier levels of anthropogenic emissions of carbon dioxide and other greenhouse gases" would be desirable.[68] The FCCC's principal mechanism for reducing emissions of greenhouse gases is an information and response system similar in form to that described in this chapter.

For developing countries, the reporting requirements are quite general and do not come into effect until three years after a country joins the regime. At that point, they are to provide a "national inventory of anthropogenic emissions by sources and removal by sinks of all greenhouse gases" and "a general description" of the steps they are taking to implement the convention. Funds are to be provided to assist them in compiling and communicating this information. Developed countries, however, are responsible for most emissions, and they have assumed extensive reporting obligations, including "a detailed description" of their policies and measures for reducing greenhouse gas emissions as well as "a specific estimate" of the effects of such policies on emissions levels, "with a view to returning individually or jointly to their 1990 levels of emissions of anthropogenic emissions of carbon dioxide and other greenhouse gases."[69] The first report is to be filed within six months after the treaty goes into effect. The reports of developed countries are to be reviewed by the Conference of the Parties at its first session and periodically thereafter.

As noted above, the secretariat's functions with respect to these reports is limited to receiving them and transmitting them to the Conference of the Parties. The Subsidiary Body for Scientific and Technological Advice, however, is empowered to "prepare scientific assessments on the effects of measures taken in the implementation of the Convention," which entails the possibility for substantive review of national reports.[70] It is also charged with assessing the general state of scientific knowledge about climate change and its effects. The Subsidiary Body for Implementation is to consider the developed country reports "in order to assist the Conference of the Parties in carrying out" its review functions and "in the preparation and implementation of its decisions."[71] Thus, the substantive functions performed by the secretariat in such international organizations as the IMF and the International Atomic Energy Agency are delegated by the FCCC to committees of the parties. If problems with implementation arise, they are to be handled by a multilateral consultative process to be established by the Conference of Parties at its first meeting. Presumably this process will be patterned after the one established under the Montreal Protocol, which provides for investigation and conciliation by a special committee and, if that fails, a report to the Conference of the Parties, which can make recommendations for corrective action. As with most treaties, environmental or otherwise, there is no provision in the FCCC for trade sanctions, or sanctions of any kind.

What seems to be envisioned is a rather anemic compliance program. There is no provision for systematic and energetic coordination of national policies and activities. Although the treaty calls for the support of national and international research programs and the development of educational and public awareness programs, these provisions are addressed to the parties, and the international organization is given no express mandate with respect to these matters. The Subsidiary Body for Science and Technological Advice has no general charge to oversee and is not empowered to coordinate national and international research on climate change or to evaluate overall climate trends. There is little indication that a vigorous effort to achieve effective and accurate reporting, or an aggressive monitoring and verification program, is contemplated. The formal commitments are too vague to permit clear-cut identification of noncompliant conduct. Responses to any deficiencies in performance are left to ad hoc decisions of the Conference of the Parties. The treaty makes provision for subsequent protocols, but there is no hint of an active program to adapt the convention to new developments or to strengthen the commitments of the parties.

It may be possible to cure these lacunae by interpretation or practice in the years ahead, but the textual basis for such a development is skimpy. Unless there is some such evolution, the prospects for success of the FCCC are doubtful.

CONCLUSION

Under the best of circumstances, several characteristics of the climate change issue are likely to lead to significant noncompliance for a considerable time. Many actors will see the current costs of undertaking needed activities as higher than the current benefits of the status quo. In the climate change arena, not only do the costs involve large and continuing outlays of funds and major changes of economic infrastructures (and even of ways of life), but the benefits are uncertain and will come far in the future. At any reasonable discount rate, the current costs may appear unconvincing as a reason for action.

Even where national governments might agree that reducing greenhouse gas emissions is a virtuous goal, they may not feel it deserves priority over the other pressing needs of their countries. This may prove as true for developed countries, such as the United States and members of the EC, concerned about jobs and economic growth as for developing countries concerned about economic development and feeding and housing their populations.

The breadth and complexity of activities that contribute to global warming and the degree to which they are interwoven with everyday behavior mean that the time required for changes to occur will be considerably longer than what ardent proponents of action would like. Achieving compliance, even by those with incentives to do so, will take time. All of these factors suggest that significant noncompliance should be expected with any international rules that establish exacting limits on behaviors that emit greenhouse gases. To recognize this fact can make expectations more realistic and help avoid the disappointment, frustration, and despair that might lead to reduced efforts to resolve these problems.

As the almost seventy-year history of oil pollution demonstrates, the development of the political will and international, national, corporate, and individual machinery to address a recognized problem can take decades.[72] In the case of oil pollution, the level of oil inputs relative to the carrying capacity of the ocean was adequate to tolerate this delay. In the case of climate change, such circumstances may not be in our favor. We may be forced to learn and adapt more quickly.

This chapter has emphasized the need to facilitate compliance, reporting, verification, and responses to noncompliance by those actors already predisposed to perform these tasks. Efforts to alter incentives and capacities are less likely to succeed than are efforts to elicit the highest possible cooperation from existing incentives and abilities, which may be making a virtue of necessity. Nonetheless, this chapter has demonstrated several ways to craft a compliance system that is better than frequently proposed alternatives. The demands of such a system will be extensive and will require nations, corporations, NGOs, and individuals to dedicate significantly greater resources to the task than they have to other environmental problems.

Several other processes can and should be set in motion to address the underlying factors inhibiting compliance for the foreseeable future. First, a major enterprise that is already under way and should continue is research on the greenhouse gas problem. Greater knowledge of the timing and nature of the effects of global warming can increase the perceived benefits of action, even if the costs of mitigation remain the same. To have this effect, these research efforts must involve a wide group of researchers and policymakers. The resultant knowledge must be made available and be assimilated widely. Successful social processes must also include evaluating and changing those modes of thinking and patterns of development that are less directly connected to climate change, including population growth.

Second, research should also be directed at developing a greater understanding of technologies and processes that can reduce greenhouse emissions. These efforts must go beyond mere development of technical fixes to include a better understanding of the factors that influence the speed and depth with which new technologies and processes are accepted in different regions of the world.

Third, an increased effort to address population growth not only would have long-term benefits in decreasing one of the major sources of global stress underlying greenhouse emissions, but also would have significant benefits in terms of development and quality of life. Without addressing this issue (and doing so with a long-term perspective), many of the other actions discussed here may prove inadequate to forestall global warming.

Finally, efforts should continue to encourage states and their citizens to adopt new perceptions of the costs of maintaining current conceptions of sovereignty. The development of the EU has provided one example of states redefining the rules of international relations on the basis of altered notions of sovereignty. Opportunities for even minor modifications of existing conceptions, such as allowing states to tax companies or to prosecute them for violations occurring within another country's borders, should not be lost. Developing new notions and definitions of sovereignty is by no means a prerequisite for achieving substantial progress toward an effective climate change regime, but modifications of current views of the sanctity of the state may merit consideration as one set of policy changes that would facilitate greater compliance with treaties and, in turn, greater mitigation of greenhouse gas emissions.

NOTES

1. See, for example, Oran Young, *Compliance and Public Authority: A Theory with International Applications* (Baltimore: Johns Hopkins University Press, 1979); Roger Fisher, *Improving Compliance with International Law* (Charlottesville: University Press of Virginia, 1981); Abram Chayes and Antonia Chayes, "Compliance without Enforcement: State Behavior under Regulatory Treaties," *Negotiation Journal* 7, no. 3 (July 1991): 311–330; Abram Chayes and Antonia Chayes, *The New Sovereignty: Compliance with International Regulatory Agreements* (Cambridge, MA: Harvard University Press, 1995); and Ronald B. Mitchell, *Intentional Oil Pollution at Sea: Environmental Policy and Treaty Compliance* (Cambridge, MA: MIT Press, 1994.

2. While the factors determining compliance will vary, corporate and individual actors can be divided into similar categories according to their predisposition toward compliance.

3. Fisher, in *Improving Compliance*, refers to "initial compliance" as "first order" compliance.

4. We use the term "compliance information system" to designate what is often referred to as "monitoring" or "verification," because these latter terms suggest an excessive focus on detecting violations. The problem often requires identifying who is complying and understanding the sources of compliance as much as knowing who is violating. Similarly, we use "compliance response system" rather than

"enforcement system" to capture the notion that appropriate responses are likely to depend more on positive action to facilitate and encourage compliance than on sanctions to deter or punish violations.

5. The contribution of a specific gas to global warming is measured on the basis of its "forcing" potential and the magnitude of current and projected emissions, that is, on the temperature increase per unit of emissions and the quantities of emissions.

6. Rules based on these criteria will lead to greater greenhouse gas reductions if and only if compliance levels are equal across activities or if higher compliance levels correlate with higher greenhouse gas contributions. Unfortunately, neither can be safely assumed.

7. Thomas A. Barthold, "Issues in the Design of Environmental Excise Taxes," *Journal of Economic Perspectives* 8, no. 1 (winter 1994): 133–151.

8. See Stram's detailed discussion of this point in his chapter in this volume. Michael Grubb also provided an extended discussion of the problems of uncertainty in quantitative outcomes under a carbon tax system in his "The Greenhouse Effect: Negotiating Targets" (*International Affairs* 66, no. 1 [January 1990]: 67–89).

9. Barthold, "Excise Taxes."

10. Paul Lewis, "Delegates at Earth Summit Plan a Watchdog Agency," *New York Times,* 7 June 1992, 20.

11. While in practice these components may become coupled, the CIS for a climate change treaty is distinct from, and should not be confused with, an environmental monitoring system, which collects information to evaluate how the environment is responding to human behavior. A CIS collects information to evaluate human behavior itself.

12. The term "transparency" has been used by Chayes and Chayes in "Compliance without Enforcement" and by Young in *Compliance and Public Authority*. The term "verification suitability" has been used with specific reference to climate change by W. Fischer, J. C. di Primio, and G. Stein in *A Convention on Greenhouse Gases: Towards the Design of a Verification System* (Julich, Germany: Forschungszentrum Julich GmbH, 1990).

13. Fischer, di Primio, and Stein, *Convention on Greenhouse Gases*, 3. See also Barthold, "Excise Taxes."

14. Of course, the fraction of global emissions of a given gas that is due to a specific source type will influence how effective such a regulation would be at reducing total emissions.

15. The treaty defines consumption as production plus imports minus exports (Richard Elliot Benedick, *Ozone Diplomacy: New Directions in Safeguarding the Planet* [Cambridge: Harvard University Press, 1991], 79–82).

16. Unfortunately, most greenhouse gas emissions arise as a by-product of otherwise beneficial economic activities and lack satisfactory substitutes, thus making bans on those activities highly unlikely. CFCs are an exception.

17. Convention on Wetlands of International Importance Especially as Waterfowl Habitat, 2 February 1971, 996 U.N.T.S. 245, 11 I.L.M. 969 (1972), 5 I.P.E. 2161 (hereinafter Wetlands Convention).

18. While equipment requirements are rightly criticized for being economically inefficient, they have the advantage of being far more readily verified.

19. For example, the International Convention for the Prevention of Pollution from Ships (2 November 1973, 12 I.L.M. 1319 [1973], 2 I.P.E. 552 [hereinafter MARPOL]) requires that signatories "ensure the provision" of reception facilities for marine pollutants but does not specify whether nations or companies must provide them.

20. The potential lack of governmental control contrasts with the situation in arms control treaties and the GATT (30 October 1947, 61 Stat. A11, 55 U.N.T.S. 187), in which the policies addressed by the intergovernmental treaty reflect largely governmental, not private, actions.

21. Fischer, di Primio, and Stein, *Convention on Greenhouse Gases*, 35.

22. Lewis, "Watchdog Agency."

23. See Fischer, di Primio, and Stein, *Convention on Greenhouse Gases;* and Abram Chayes and Eugene B. Skolnikoff, "A Prompt Start: Implementing the Framework Convention on Climate Change" (unpublished paper, Cambridge, MA, 28–30 January 1992), 5.

24. Fischer, di Primio, and Stein (*Convention on Greenhouse Gases*, 20) described the development of a standard questionnaire as an "urgent task," but did not address how reporting can be encouraged.

25. See, for example, U.S. General Accounting Office, *International Environment: International Agreements Are Not Well Monitored*, GAO/RCED-92-43 (Washington, DC: U.S. GPO, 1992); Gerard Peet's report, "Operational Discharges from Ships: An Evaluation of the Discharge Provisions of the MARPOL Convention by Its Contracting Parties" (Amsterdam: AIDEnvironment, 15 January 1992); and Ronald B. Mitchell, *Intentional Oil Pollution at Sea*, chapter 4.

26. The almost perfect reporting records of the International Convention for the Regulation of Whaling (2 December 1946, T.I.A.S. No. 1849,

161 U.N.T.S. 72) and the Memorandum of Understanding on Ports State Control in Implementing Agreements on Maritime Safety and Protection of the Marine Environment (26 January 1982, 21 I.L.M. 1 [1982], I.P.E. II/A/26-01-82) demonstrate the willingness of states to report when they perceive reporting as essential to accomplishing the goals of the agreement.

27. An example of the failure to request information explicitly is evident in the International Maritime Organization's standardized reporting format for the MARPOL Convention, which does not include a space for the report period, making determination of the report period difficult in many cases.

28. The EC reports ozone-depleting substance emissions aggregated not only across gases but also across countries, but this practice makes it impossible to determine whether individual states are in compliance (Benedick, *Ozone Diplomacy*, 181).

29. Fischer, di Primio, and Stein, *Convention on Greenhouse Gases*, 16.

30. Gerard Peet, in "Discharges from Ships," came to a similar conclusion with respect to reporting under the MARPOL Convention.

31. Fischer, di Primio, and Stein, in *Convention on Greenhouse Gases*, suggested that member states should create national authorities responsible for acquiring, aggregating, and reporting "national emissions or other convention-relevant activities by the country" (p. 7).

32. The Memorandum of Understanding on Port State Control has developed a computerized system (Secretariat of the Memorandum of Understanding on Port State Control, *Annual Report 1990* [The Hague: The Netherlands GPO, 1990]).

33. Fischer, di Primio, and Stein, *Convention on Greenhouse Gases*.

34. Montreal Protocol on Substances that Deplete the Ozone Layer (16 September 1987, 26 I.L.M. 1541 [1987], Art. 10).

35. Fischer, di Primio, and Stein, *Convention on Greenhouse Gases*, 35.

36. Convention on Long-Range Transboundary Air Pollution (13 November 1979, 18 I.L.M. 1442 [1979]).

37. See Marc Levy, "European Acid Rain: The Power of Tote-Board Diplomacy," in *Institutions for the Earth: Sources of Effective International Environmental Protection*, eds. Peter Haas, Robert O. Keohane, and Marc Levy (Cambridge: MIT Press, 1993). For a description of a similar system used for analysis of trade statistics in the Convention on International Trade in Endangered Species, see Mark C. Trexler, "The Convention on International Trade in Endangered Species of Wild Fauna and Flora: Political or Conservation Success?" (Ph.D.

diss., University of California at Berkeley, 1989); and U.S. General Accounting Office, *International Environment.*

38. Lewis, "Watchdog Agency."

39. Daniel Navid, secretary-general, Wetlands Convention, interview, 1991.

40. In arms control, these methods are known as national technical means, or NTM.

41. Countries with satellite capabilities include the United States (military and LANDSAT), Russia, France (SPOT), the EU (ESA), Japan, Brazil, and India (Fischer, di Primio, and Stein, *Convention on Greenhouse Gases,* 22).

42. Ibid., 22–26.

43. Chayes and Skolnikoff, "A Prompt Start," 11–12.

44. Comparability involves converting data into standard currencies and adjusting for inflation, for example.

45. Timothy Egan, "Space Photos Show Forests in Pacific Northwest Are in Peril, Scientists Say," *New York Times,* 11 June 1992, A13.

46. While the two strategies are logically distinct, in practice they may often be coincident, as when continued financing for a project acts as a positive incentive for compliance and the elimination of such financing acts as a sanction for violation.

47. Of course, nations will also need to determine whether and how they will respond to actors who fail to meet the requirements established in domestic treaty-implementing legislation. Nations will choose a wide variety of approaches to encourage compliance and discourage violation based on sociopolitical factors as well as on the effectiveness and efficiency of the approach. For brevity's sake, we do not discuss the various advantages and disadvantages of different domestic-level response strategies.

48. Hilary F. French, "After the Earth Summit: The Future of Environmental Governance," Worldwatch Paper 107 (Washington, DC: Worldwatch Institute, 1992), 31.

49. French, "After the Earth Summit," 31.

50. Lewis, "Watchdog Agency."

51. See Timothy B. Hamlin, "Debt-for-Nature Swaps: A New Strategy for Protecting Environmental Interests in Developing Nations," *Ecology Law Quarterly* 16, no. 4 (1989): 1065–1088.

52. For historical examples, see George F. Kennan, "To Prevent a World Wasteland: A Proposal," *Foreign Affairs* 48, no. 2 (April 1970): 401–413; and Grenville Clark and Louis Sohn, *World Peace through*

World Law (Cambridge: Harvard University Press, 1960). More recently, see French, "After the Earth Summit."

53. See, for example, Grubb, "Greenhouse Effect."

54. Lewis, "Watchdog Agency."

55. Peter Haas, "Protecting the Baltic and North Seas," in Haas, Keohane, and Levy, *Institutions for the Earth.*

56. The International Labor Organization uses blacklisting extensively to encourage reporting.

57. For example, see the Earth Island Institute's advertisement urging trade sanctions against Norway for its resumption of whale hunting (*New York Times*, 3 May 1993, A15).

58. See, for example, Secretariat of the Memorandum of Understanding, *Annual Report 1990*. George Kasoulides has discussed this problem with the Memorandum of Understanding at length ("Paris Memorandum of Understanding: A Regional Regime of Enforcement," *International Journal of Estuarine and Coastal Law* 5:1-3 [February 1990]: 180–192).

59. See Barthold, "Excise Taxes"; and U.S. Congressional Budget Office, *Carbon Charges as a Response to Global Warming: The Effects of Taxing Fossil Fuels* (Washington, DC: U.S. GPO, 1990).

60. Grubb, "Greenhouse Effect."

61. French, "After the Earth Summit," 30.

62. James K. Sebenius, "Negotiating a Regime to Control Global Warming," in *Greenhouse Warming: Negotiating a Global Regime* (Washington, DC: World Resources Institute, 1991), 36.

63. See, for example, Gary Clyde Hufbauer and Jeffrey J. Schott, *Economic Sanctions Reconsidered: History and Current Policy* (Washington, DC: Institute for International Economics, 1985).

64. See Ronald B. Mitchell, "Intentional Oil Pollution of the Oceans," in Haas, Keohane, and Levy, *Institutions for the Earth;* and Mitchell, *Intentional Oil Pollution at Sea.*

65. Gene S. Martin, Jr., and James W. Brennan, "Enforcing the International Convention for the Regulation of Whaling: The Pelly and Packwood-Magnuson Amendments," *Denver Journal of International Law and Policy* 17, no. 2 (winter 1989): 293–315; and Dean M. Wilkinson, "The Use of Domestic Measures to Enforce International Whaling Agreements: A Critical Perspective," *Denver Journal of International Law and Policy* 17, no. 2 (winter 1989): 271–292.

66. French, "After the Earth Summit," 30.

67. For an extended analysis of compliance under the MARPOL Convention, see Mitchell, *Intentional Oil Pollution at Sea.*

68. UN Framework Convention on Climate Change (15 May 1992, UN doc. A/AC.237/18 [Part II]/Add.1), Art. 4(2).

69. Ibid., Art. 12.

70. Ibid., Art. 9(2).

71. Ibid., Art. 10.

72. See Mitchell, *Intentional Oil Pollution at Sea.*

5 The Triad as Policymakers

■ **RAYMOND VERNON**
John F. Kennedy School of Government
Harvard University

Some of the world's most critical problems of environmental degradation can only be effectively tackled through international agreements covering the world's major countries. This proposition is universally accepted by policymakers on the environment and is echoed in some of the chapters in this volume. It explains why policymakers and academics devote so much time and effort to defining the terms of the required international agreements.[1] In that context, the following is a list, only partial, of questions that need to be addressed: Should agreements be mandatory or hortatory? Should they be rule-based or result-based? Should they be global or regional? Should monitoring and enforcement be supported by a formal organization, and if so, what kind of organization?

To answer such questions with some assurance, however, one must understand the internal policy-making processes of the negotiating states themselves, processes that differ markedly among countries in response to their distinctive histories, values, and institutions. In this chapter, I examine the national policy-making processes of three influential players in any upcoming negotiations on the environment: the United States, Japan, and the EC (now the EU), which make up the triad from which the chapter takes its title. Other countries, of course, are also bound to influence the content of future agreements, including Brazil, India, and China, and scholars intent on shaping the terms of such agreements would do well to analyze their decision-making processes as well. This chapter, therefore, only contributes to the analysis of internal processes as they may eventually affect international agreements.

In each of the three cases that follow, I draw on the extensive literature on the behavior of the three players as they have formulated policies affecting the distribution of costs and benefits within their respective economies. That literature demonstrates the power and persistence of the national characteristics that are likely to distinguish the players' roles in environmental negotiations.

I conclude with a note on a characteristic that appears to play a special role in environmental issues: grassroots responses arising outside of established policy-making channels. Operating through transborder alliances, grassroots organizations appear to have exerted an unusual degree of influence on the negotiation and adoption of new initiatives. They are, however, likely to be much less effective in the day-to-day implementation of international agreements, an activity in which the established decision-making apparatus of government will probably dominate. On issues of implementation, therefore, an understanding of the history, institutions, and values of participating governments is likely to prove especially important.

U.S. PATTERNS

Generalizations about national roles in international negotiations have to be made with caution, as there is always at least one memorable case that will not fit the posited pattern. However, there are strong grounds for the generalization that U.S. representatives are more prone than are their counterparts in most other countries to take an activist role in the process: U.S. negotiators tend to place new propositions on the table, to organize blocking coalitions, and to modify or even reverse positions in the course of the negotiation.

It is doubtful that the propensity of U.S. negotiators to take an activist line is due to either greater wisdom or energy. Instead, part of the reason probably lies in a persistent need on the part of the country's negotiators to justify the country's claimed position as the leader of a Western coalition. In addition, U.S. representatives could not respond with the seeming activism that often characterizes their international negotiating style if it were not for certain characteristics of the country's internal decision-making processes. By comparison with most countries, that process usually tolerates a relatively high degree of initiative and flexibility on the part of U.S. representatives.

THE SEPARATION OF POWERS

It may seem paradoxical that representatives of a country whose government is built on an elaborate system of checks and balances should appear to have a relatively high degree of flexibility in international negotiations. But the indispensable requirement for an effective system of checks and balances is the separation of powers of the various governmental entities involved; within certain limits, each entity is free to act without securing the consent of the others in the system. Although the executive branch may consult congressional leaders regarding its positions in upcoming international negotiations, it will stop short of asking for advance congressional approval; and in the unlikely event that it asked for such approval, it would surely be rebuffed.

The flexibility of U.S. representatives in international negotiations is often enhanced by the fact that the principle of separation of powers creates walls not only between the three main branches of government but also within the executive branch among its departments and commissions. Agencies such as the Securities and Exchange Commission, the Federal Reserve Board, and the EPA look on their powers as being determined by statute, not by the White House. Although ordinarily deferential to White House views, these agencies are likely to defend their turf against any challengers. Their tolerance for the views of other agencies in the executive branch that are affected by the exercise of their statutory powers is often fairly limited.[2]

THE REVOLVING DOOR

The probability that affected agencies will be slow to consult with one another over their common problems is also due to the nature of the leadership of each administration. With each change in administration, a new group of about three thousand officials is brought to Washington, recruited from the four corners of the country, trained in a variety of professions, and sharing little in background and values.[3] Exempted from civil service standards and appointment processes, all that the appointees can be expected to have in common is political credentials that have survived a screening conducted by the party occupying the White House. Before taking up their posts, individuals in the group characteristically have had only limited contact either with one another or with the agency to which they are assigned. In addition, few of them will have any expectation of building a career in the federal bureaucracy.[4]

These generalizations, of course, have been stronger for some administrations and for some agencies than for others. They were stronger, for instance, in the Reagan era than in the Bush administration; and with the turnover in top personnel attending the shift from a Republican to a Democratic president, they gained in strength again in the Clinton administration. While some of the political appointees involved in this revolving-door process may take their posts with deep personal commitments to some given line of policy, the prospect of a relatively short tenure places a high premium on making an early mark with little regard for consistency or continuity.

For all these reasons, U.S. policymakers are likely to frame their objectives and shape their tactics for selling their proposed policies without much hope or expectation of developing a genuine consensus among the agencies that have a stake in the issue. Interagency committees may exist in profusion. But the persistent tendency of the policymaker will be to defend the power to operate autonomously and, if that option is not available, to look for a strategy, such as enlisting the president's personal support, that will allow the policymaker to finesse the process of consultation inside the executive branch.[5]

In a national decision-making structure in which consensus does not dominate, a major element of unpredictability is introduced in the positions that negotiators are likely to take in international circles. The history and tradition of each agency will prove an uncertain predictor of its position on any new issue. A great deal will depend on the personal motivations and the bureaucratic skills of the agency's top echelons. An agency head charged with a particular functional area such as the environment may not have the motivation or skill to overcome a blocking element within the White House staff. In contrast, an agency head with an inside track to the White House will often be able to introduce proposals that represent abrupt departures from past policies.

The mercurial role of the executive branch of the U.S. government in environmental policies has been strikingly apparent in the Carter, Reagan, and Bush administrations, and characteristic of the Clinton administration as well. Some of the sharp reversals in policy between the Carter and Reagan administrations, such as U.S. abandonment of support for restrictions on the international movement of hazardous wastes, obviously represented a shift in the personal philosophy of the president; it may even have mirrored a small shift in the national consensus on such matters. But the zigzag course of the executive branch in the latter 1980s and in the 1990s seemed much more a product of tactical shifts in domestic politics, coupled with shifts in the preferences and objectives of a few key policy entrepreneurs in the Carter, Reagan, and Bush administrations.[6]

THE ROLE OF CONGRESS
Recognizing the principle of the separation of powers and the independence of the three federal branches of government, congressional leaders

will not ordinarily question the right of the executive branch to launch any proposition for international discussion, as long as Congress retains the opportunity eventually to pass on its merits. Sometimes, individual members of Congress may grouse at the executive branch's exercise of that discretion, fearing that the very introduction of a proposal in an international forum could tie the legislature's hands at a later stage. At other times, however, members of Congress will be much relieved at not being required to take an early position on some contentious issue.

Moreover, the U.S. executive branch itself, when conducting an international negotiation over an economic policy, has been known at times to regard the independence of Congress as a negotiating advantage, because it allows U.S. negotiators to threaten the representatives of other countries with the possibility of congressional displeasure and retribution if they do not accept the proposals offered by the executive branch.[7]

But the congressional drive to retain power and independence means that the texts of laws and regulations take on special importance, becoming the centerpiece in struggles with the executive branch over national policies. Where the texts of laws and regulations are inexplicit in prescribing the standards or procedures to be followed, the discretion of the executive grows at the expense of the legislature. The emphasis on standards and procedures, in turn, elevates the role of adversarial proceedings and formal process, places the legal profession at the center of the controversy, and thus limits the power of the bureaucracy to make complex judgments and to rely on inexplicit criteria.

EFFORTS TO COPE

As numerous observers have pointed out, any international negotiation conducted among democratic governments is bound to proceed at two levels simultaneously, one involving the interaction among governments, the other involving the interaction of each government with its domestic interests. In the case of the United States, innovators commonly launch their proposals in international settings without first developing a broad national consensus. As a result, there is a high risk that the innovator may not be able to retain U.S. support for the position originally advanced. Some of the interests overlooked in the first phase of the negotiating process will be eager to make their position felt in subsequent rounds, and in some instances, their late intervention may carry such weight as to require major changes in the U.S. position.[8] The porousness of the U.S. decision-making structure represents an open invitation to any such neglected interest. If the opposition has not already captured a sympathetic ear in one agency, it may be able to do so in another.[9] And if the executive branch is unmovable, the Congress, the courts, or the media may offer an alternative channel.

The system, therefore, places a premium on aggressive advocacy in any national debate, a characteristic especially in evidence in the formulation of environmental policies. Sheila Jasonoff describes the process as "confrontational, litigious, formal, and unusually open to participation."[10] In such a setting, the possibility that some group may force a revision of a U.S. position in an international negotiation is relatively high. This characteristic adds to the risk that the United States may shift course abruptly in the negotiation of an international agreement and may find itself obliged to breach agreements after they have been adopted.

There have been times in years past when the executive branch, in an effort to clear the path for an international negotiation, has tried to bypass Congress, claiming already to have the power to enter into agreements in the name of the United States. However, experience suggests that such an approach is highly vulnerable, especially in a policy area that may involve special interests and may require changes from time to time in U.S. legislation. Executive agreements not expressly authorized or approved by Congress run the risk of being ostentatiously disregarded on Capitol Hill. For example, Congress has been cavalier about U.S. violations of its commitments under the GATT, sometimes ignoring the existence of the violations for long periods, sometimes grudgingly making an adjustment in response.[11]

Moreover, there have been signs that the advantages which U.S. representatives derive from their autonomy in launching international negotiations are declining over time. Other countries have begun to realize that the proposals of U.S. representatives may not be backed by a broad U.S. consensus and that adversaries inside the United States may eventually force major alterations in a brave U.S. initiative. Indeed, this scenario has characterized the history of international negotiations in matters of foreign trade: by the late 1970s, many countries were resistant to undertake any negotiations with the U.S. executive branch for fear that the results would not be acceptable to the U.S. Congress. To continue with such negotiations, some remedy had to be found.

The remedy that was fashioned for the conduct of trade negotiations, the so-called fast-track provisions, provides a precedent that could be extended to other areas. Under the fast-track provisions, the executive branch agrees to conduct its negotiations in close consultation with private interest groups and with selected congressional representatives. In return, Congress agrees that it will vote the negotiated agreement up or down, without delay and without qualification or amendment. In practice, there are also other devices that sometimes allow the executive to engage in international negotiations with reasonable assurance that Congress will not block the resulting agreement. For instance, while operating under congressional dispensations, various agencies in the executive branch are empowered to negotiate Memoranda of Understanding with foreign countries in a number of different fields.

Even when the executive has acted well within its authority, however, there still may be a risk that Congress or the courts, responding to the initiatives of special interests, will compel the U.S. government to act in disregard of existing international commitments. Meeting this challenge may appear to be equivalent to asking the oceans to stand still. However, plenty of evidence suggests that members of Congress themselves are often looking for some insulation against the unremitting pressures of special interests. Some of their efforts in that direction have had unfortunate consequences for the international negotiation process. In trade negotiations, for example, in order to appear responsive to special interests without getting entangled in their individual cases, Congress has commonly enacted provisions that increase the ability of special interest groups to put pressure on the executive branch. These provisions typically open up new avenues of petition to executive agencies or the courts, lay down explicit standards that are supposed to guide the executive agencies and the courts toward a decision, and so forth.[12]

Still, on the whole, the executive branch and Congress often share a desire to limit their exposure to special interest pressures in individual cases, provided that they can find a way which is not too costly in political terms. This sentiment opens up the possibility that international agreements which include compulsory arbitration clauses may sometimes be a welcome escape for both, offering a way of depoliticizing the handling of individual cases.

The United States and Canada took a large step in that direction in the negotiation of the U.S.-Canada free trade agreement, with the formation of a binational court of appeals to deal with issues involving subsidies and dumping and with the extension of compulsory arbitration panels to deal with disputes over other issues covered in the treaty. Such innovations could well point the way to the future structuring of other international agreements.

In the U.S. case, one can anticipate a considerable degree of activism on the part of U.S. negotiators in the course of developing new international initiatives, which will be interrupted by frequent midcourse corrections and even reversals as new influences are brought to bear on the U.S. position. In the enforcement of any agreed international standards, one can anticipate a constant struggle to ensure that the intent of the agreement is reflected in its application to individual cases.

JAPANESE PATTERNS

By ordinary standards, Japan's economy would be described as market-based and its political processes as democratic. These standards allow plenty of room for variety, and Japan's approach to the making and enforcement of public programs such as the protection of the environment includes a number of distinctive features. Although scholars do not agree in all particulars on the critical factors that shape Japanese behavior, the various analyses of that behavior share a good deal of common ground.[13]

THE DECISION-MAKING STRUCTURE

The us-versus-them syndrome, so pervasive in the formulation of the foreign policies of most countries, has been especially strong in most of the history of modern Japan. This attitude is hardly surprising, given the fact that for hundreds of years before the opening of Japan in 1868 the country was in peril of becoming a prize for the prowling navies of the Western powers.

During the century following the opening of Japan, the country's leaders struggled bitterly among themselves on many issues, but they remained remarkably united in their view of the paramount domestic objectives. In 1868 it seemed obvious to the Japanese elite that their national existence depended on their ability to absorb the war-making technology of the West. Once that objective was within reach, by about the close of World War I, the Japanese elite struggled to solve another seeming threat to their national existence—their utter dependence on imports of raw materials. Repeated efforts on the part of Japanese industry to gain direct access to sources of oil and minerals were rebuffed by strong cartels composed of leading European and North American firms.[14] Struggling to overcome these vulnerabilities, Japan eventually invaded China and attacked Pearl Harbor. During this phase, there were occasional signs that the unanimity among Japan's leaders that had apparently prevailed in earlier decades was not quite as complete, but deviants were quickly brought into line by political assassinations and other forms of pressure.

With Japan's defeat in World War II, national priorities changed. Economic recovery became the consuming objective of the economy. Once again, it would have been hard to find any part of Japanese leadership prepared to subordinate that objective to some other purpose.

It was not until the 1980s that one could see any significant modifications in the us-versus-them approach or any significant measure of recognition in Japan that the country had a major stake in solving some problems that lay beyond its own borders. An indication of the extent of that change was Japan's willingness in 1987 to back the U.S.-inspired proposal for an international agreement to curb the use of CFCs.[15]

Japan's tendency to present an unaccommodating united front to the rest of the world, however, has been due to factors that cannot be expected to change very rapidly. Paradoxically, one such characteristic has been the strength, stability, and professionalism of the Japanese bureaucracy. With lifetime commitments to their profession, bureaucrats engaged in the policy-making process characteristically have operated in a setting in which the identity of the principal players was highly predictable over extended periods of time. In such a setting, game theorists remind us, players are encouraged to develop a reputation for team playing, albeit team playing tempered by toughness. Some of the characteristics that are typical of U.S. bureaucratic behavior, therefore, are frowned upon in Japan, including opportunistic innovation coupled with tactics that take advantage of the absence or temporary weakness of the opposition.

The stability of Japan's bureaucracy has been matched by the stability of other elite sectors of Japan's decision-making structure. A single political party, the Liberal Democratic Party (LDP), dominated the government from the end of World War II until the 1990s, and the hold of that party seemed unshakable. True, the party was always strained by internal rivalries among its so-called factions, each headed by a prominent politician eager for the prime minister's office, but that rivalry was not strong enough to threaten the party's hold on the government.

In the business world, the degree of the leadership's stability during the postwar period also has been remarkable, given the dimensions of the growth and change in the economy. The predictability and stability of the leadership have been bolstered by the lifetime career patterns of Japanese business executives, by the links among business firms created through exchanges of shares and other long-term arrangements, and by the dominant position of Japan's peak business organizations, notably the Keidanren. With so stable an elite in control, the disposition of policymakers for long-term reputation building and team playing has been high.

Such stability, of course, could not continue indefinitely. In 1993, it was at last brought to an end. In a similar vein, the stability and predictability of the career patterns of Japan's business executives were being disturbed by a sharp increase in job-hopping and an increase in demands for leisure time.[16] However, it is not yet clear if these countertrends portend any gross change in the patterns of Japan's economic decision making.

Until the 1990s, the striking absence of innovation or initiative on the part of the Japanese was apparent with regard to practically all international issues, not just environmental ones. My explanation for this pervasive Japanese characteristic is the mirror image of my explanation for the opposite propensities of U.S. representatives: Innovative proposals usually entail a giant step into the unknown, with latent risks and uncertain benefits for some groups in the population. In a move based partly on faith even when justified by reason, innovators typically are obliged to ride roughshod over the doubts and misgivings of some groups at home likely to be affected by the proposed policy. In the U.S. system of decision making, with consensus infeasible and artful dodging a normal part of the game, it is sometimes possible to launch an innovative proposal that embraces some general principle, despite the existence of such unassuaged misgivings. In the Japanese system, with every major interest in a position to bring the negotiation to a standstill, the possibility of obtaining agreement to enter into unknown areas by espousing a general principle is greatly reduced. Japan's continued unwillingness to agree to a ban on whale fishing in international negotiations represents one more illustration of its inability to ride roughshod over unreconciled opposition.

The same distinction helps to explain another characteristic of Japanese decisions: foreign pressure, *gaiatsu*, appears to play a very important role in determining Japanese moves, especially where the breadth of the subject matter engages the interests of a number of ministries. However, history suggests that the use of foreign pressure has its limits. If a significant part of the Japanese establishment feels that such pressure has been excessive and that a genuine internal consensus does not exist, it may well take an independent course in spite of a contrary international agreement. *Gaiatsu* may be effective, however, when it provides Japanese negotiators with a way out of their internal dilemma of building a reputation as both tough bargainers and team players. To maintain such a position, the participants must see to it that the concessions made to achieve internal agreements are hard-won and small in scope. The critical role of foreign pressure may be explained by the fact that it gives a resisting group an excuse for yielding without losing its reputation as a tough bargainer.

Finally, one other characteristic of Japanese decision making seems to flow from this line of speculation. In contrast to the United States, the course of Japanese policies partakes of some of the characteristics of a supertanker under way. The probability that some entity in the government structure will take an independent line and disregard or override an existing international agreement seems low. Inertia and momentum play a dominant role. Any future change in direction, as with the changes of the past, will only be effected slowly and with great effort.

ENVIRONMENTAL POLICIES

Japan's history of adopting national environmental policies has deviated sufficiently from its behavior in other policy fields to raise questions about Japan's likely future role at the international level. In a country that places a high value on the avoidance of confrontational tactics, militant action of self-empowered environmental groups has been largely responsible for placing environmental issues on the national agenda. Although it is not my purpose to review Japan's domestic environmental policies in this brief chapter, it is useful to recall a few highlights of the Japanese record simply to test the extent to which the country's characteristic decision-making processes are applicable to its handling of environmental issues.[17]

Because of the high population densities on Japan's main islands and the spectacular growth of its industrial facilities in the 1950s and 1960s, Japan was one of the first industrialized market economies to react to some of the acute effects of modern industrial pollution. By the 1960s, the Japanese were discovering widespread instances of poisoning from toxic metals and chemicals, including the notorious *minimata* and *itai-itai* epidemics, as well as the mushrooming of bronchial asthma cases in some cities. However, the policy-making machinery, resting securely in the hands of the bureaucracy, the Keidanren, and the dominant LDP, seemed insulated from the local pressures of ordinary citizens. As long as these developments were confined to limited areas of the country and as long as they did not interfere with the achievement of the Japanese government's dominant objectives of rapid, sustained growth, the groups that were affected seemed to have no choice but to bear the pollution's effects.

Japanese environmental conditions as a whole were probably no worse than those of Western Europe or North America; indeed, if life expectancy data are any guide, they may even have been better. The consequences to the affected local areas in Japan, however, were so shocking and the threat of further consequences so unremitting that local groups searched in desperation for some remedy. One feature of their response was thoroughly Japanese: they organized themselves in local cooperative groups that unrelentingly pushed local polluters and local government officials to develop an appropriate response. Another feature was a remarkable aberration, presumably brought on by the gravity of the situation. Local groups instituted civil suits in the courts to establish the responsibility of the polluters and to obtain compensatory damages for those affected. The success of the plaintiffs in four celebrated cases created a landmark in Japanese environmental policy.[18]

By the 1970s, environmental issues in Japan were no longer the concern of local groups alone. By that time, Japan's mass media had thrown their weight in favor of environmental controls, stressing the continuation of environmental degradation and the foot-dragging and indifference of Japan's leading polluters. There were some years of shadowboxing in which the bureaucracy in Tokyo gave lip service to environmental objectives while public opinion was still being formed. But by the mid-1970s, key ministries in Tokyo, notably including the Ministry of Health and Welfare, appear to have joined the media in an alliance against industrial sources of pollution. With that shift, a new internal balance was achieved in Japan. Politicians and industrial leaders, taking note of the shift, joined in the framing of a new set of environmental policies.[19]

Unlike in the United States, however, the transformation of public opinion in Japan from indifference to commitment was not accompanied by the development of strong national organizations or by the appearance of organized pressure groups in Tokyo. The struggle that eventually produced a shift in national policy appears to have been conducted by more ephemeral means, including local movements and the media, which eventually led policymakers to recognize that the public had developed new expectations with regard to environmental policy. When that change in expectations was recognized by the bureaucracy, the LDP, and the Keidanren, the stage was set for concerted national action.

The Japanese programs of environmental control that emerged by the latter 1970s were exemplary in their stated goals and their initial achievements; the imposition by the Ministry of International Trade and Industry (MITI) of a ban on lead in gasoline posed a sharp contrast to the dreary trail of suits and countersuits over the same issue in the United States, where adversarial proceedings and statutory schedules were a normal part of the implementation process. In added contrast to U.S. practice, monetary support for those who were injured by the new restrictions played a major role in Japanese official programs.[20]

In addition, in characteristic Japanese fashion, the paper trail created by Japanese programs contained little in the way of hard commitments or unambiguous standards. The effective guidelines, if they existed at all, were contained in the side deals and confidential memorandum exchanged inside the bureaucracy.[21] Moreover, the affected industry groups were not regarded as adversaries in the process that shaped the relevant programs. On the contrary, their consultations with the bureaucracy were frequent and extensive. Finally, exercising the broad discretion that typically resides in the Japanese bureaucracy, governmental officials were both flexible and supportive in their relations with the affected firms during the implementation phase, adjusting schedules and providing for special financing as required.

By the 1990s, however, the Japanese apparently had achieved enough progress in environmental issues on the home front that such issues had slipped off the front pages of Japanese media. There were substantial indications that the policies adopted by the government were not being neglected by the bureaucracy and were being implemented with some care and efficiency. In sharp contrast to the United States, however, the implementation of these programs was being shaped much more by expert opinion than by political pressure.[22] Indeed, despite Japan's domestic activism on environmental matters, there had been no accompanying development of national movements devoted to preserving the environment, such as the Greens in Europe or the Sierra Club and the Natural Resources Defense Council in the United States.[23] Nor have dedicated environmentalists often been included in the numerous advisory groups from which the Japanese bureaucracy draws guidance. The crusading elements so evident in European and U.S. environmental movements, therefore, have been much less prominent in Japan's policymaking establishment. On the contrary, although the execution of existing policies appears notably efficient by any comparative standard, the commitment of the establishment to environmental issues continues to be largely reactive.

Japan, it appears, cannot be expected to contribute much to new international initiatives on the environment. Nor can it be expected, as a rule, to favor agreements that prescribe behavior as distinguished from those that define goals. Once committed, however, the international community can count on a level of execution from Japan that will be relatively high by international standards.

EUROPEAN PATTERNS

The case of the EU differs from that of the United States and Japan in a number of major respects. Two important differences are that the EU is a supranational institution which shares some of its responsibilities with its member states and that the responsibilities of the EU are still in the process of being formed. Still, some patterns already exist, suggesting some of the characteristics of the community's future role.

FROM NATIONAL POLICIES TO SHARED RESPONSIBILITIES

European countries enter the present era of international activity in environmental controls with a rich record, covering several decades of programs at the national level.[24] Perhaps in part because of the novelty of the issues, these programs have evolved from a great variety of approaches. With time, national programs began to show more similarities in approach, pushed in common directions by the prodding of the former EC.[25] For instance, the polluter-pays principle gradually secured greater currency. However, major national differences have persisted into the 1990s, reflecting substantial variations in public opinion over the importance of environmental concerns, along with a wide variety of legal traditions and regulatory practices.

Still, the Europeans as a group can be distinguished from the United States in a number of explicit respects, as they have exhibited an overall pattern in their approach to environmental controls that is much more akin to that of the Japanese than to that of the Americans. Like the Japanese, Europeans are far less confrontational than are Americans in the processes by which they formulate standards and enforce them. Even Germany, which has a highly vocal environmental lobby and is identified with strong environmental policies, nevertheless relies heavily on informal consultations and voluntary agreements.[26]

Contrasts between European and U.S. practice are especially evident in the processes by which scientific opinion is amassed. In the United States, the evidence on which the government formally relies is accumulated largely through overt adversarial procedures, a process that usually generates a disparate set of conclusions purportedly supported by scientific authority and objectivity. By contrast, Europe's administrators garner their evidence, to the extent that it figures in their decisions, from scientific sources largely of their own choosing.[27] What is more, both in the process by which standards are formulated and in the means by which they are enforced, Europeans tend to rely much more than Americans on private consultation with the interests most affected. Also, the responses of Europeans to noncompliance, as with the responses of the Japanese, turn out to be far more flexible than the heavy-handed U.S. approach, which makes use of fines and court orders.[28]

Although the EU's role in the environmental field has been influenced a little by the practices of its member states, its unique treaty provisions and decision-making institutions have cast it in a very distinctive role. To study the decision-making policies of the EU, one must close in on a rapidly moving target. Barely three decades old, the EC, and now the EU, is still in the process of defining its goals, shaping its institutions, and settling on its patterns of governance.

The original goals of the EC, as defined in the Treaty of Rome that created it in 1958, made no mention of the environment. According to the treaty's text, its purpose was to establish the foundations of an ever closer union among the European peoples by establishing a common market among its member states. Under the terms of the treaty, its member states agreed to surrender their powers over the cross-border movement of goods, services, money, enterprises, and workers to the institutions of the EC, with a view of creating such a common market. They agreed as well that the law-making body of the EC, its Council of Ministers, could enact the laws required to create the common market by a qualified majority vote rather than by the unanimous consent of its members.

The first phase in the evolution of the EC, from 1958 to 1966, was one in which its member countries belatedly discovered just how deeply the extraordinary commitments of the Treaty of Rome would limit their capacity for independent national action. Suddenly, for instance, France found itself unable to bar a foreign-owned enterprise such as General Motors from setting up a plant in its jurisdiction without considering the possibility that the plant might settle in Brussels and export its output to France from its Belgian location. The dawning realization that the treaty's commitments created powerful restraints on the capacity of its member states to act autonomously eventually produced a reaction from France's President Charles de Gaulle. Despite the treaty's provisions to the contrary, he exacted a commitment from member states that issues which any member country defined as vital to its interests could only be settled by unanimous agreement.

That early decision seemed to diminish the prospects of the EC becoming a significant actor in the economies of its member states, much less a force in the environmental field. But the vitality and persistence of other institutions of the EC, notably its Court of Justice, eventually opened up new possibilities. Despite the agreement that the Council of Ministers would rule by the principle of unanimity, the court still had to discharge its responsibilities as the ultimate interpreter of the meaning of the treaty. It had to rule, for instance, on complaints by individuals, national governments, and even the institutions of the EC itself regarding the legality of actions taken in light of the treaty's provisions. As it turned out, such complaints arose often and covered a wide range of issues. The executive arm of the EC, for instance, brought frequent actions against national governments, alleging that they were not living up to their treaty commitments, while individual enterprises commonly brought actions against the executive arm of the Union, the European Commission, asserting that it had used its executive power in a manner inconsistent with treaty provisions.

This grist for the judicial mill produced a series of decisions during the years from 1966 to 1985 that solidified and amplified the powers of the EC. During that period, the Court of Justice affirmed that the provisions of the Treaty of Rome and the regulations and directives enacted by its institutions to create the common market had "direct effect" on the member states—that is, they had the quality of national law and they superseded existing national law where differences existed. The court also clarified the fact that, in areas of policy in which the EC exercised its powers under the treaty, the EC also had responsibility for negotiations with outside countries.[29]

There was one crucial area, however, in which the EC's powers were limited and would remain so indefinitely. The enforcement of the EC's decisions was left largely to the member states. To be sure, over the years the EC would develop various devices to reduce the extent of noncompliance, such as requirements for reports on the implementation of its directives and recommendations, and appeals to the Court of Justice to find member states in violation of their treaty obligations. Eventually, the Court of Justice held that in some circumstances individuals could sue a member state for damages incurred by the state's failure to implement an EC directive. The EC's limited powers of enforcement, however, is nonetheless a major factor in any assessment of its role.

Despite that limitation, the EC's powers were vastly strengthened by Court of Justice decisions in a process that for a long time garnered little public attention. A small group of lawyers and scholars specializing in EC law followed these developments with avid interest; but until the 1980s, these critical developments went almost unnoticed among politicians, political scientists, and economists, not to mention the public media and the public at large.

By the early 1970s, it was beginning to be apparent that the EC could not forever remain aloof from the problems of environmental controls. Its mandate to create a common market meant that it would eventually be required to address any major obstacles that might stand in the way. On the other hand, the EC's authority to deal with barriers in the form of national environmental controls was not clear. It was evident that in some circumstances the programs of individual states aimed at controlling their national environments could have the incidental effect of blocking cross-border movements inside the common market. One illustration is the efforts of some countries to curb the sale and use of motor vehicles that did not adhere to specified emissions standards. On the other hand, the treaty authorized member states to limit exports and imports designed to protect the life and health of humans, animals, and plants, so long as they were not "a disguised restriction on trade."[30]

Despite the ambiguities surrounding the EC's powers under the Treaty of Rome in the field of the environment, in 1972 its member states authorized the EC to take a vigorous role in environmental controls. Much of the initiative for this development came from member states with relatively high environmental standards, such as Germany and the Netherlands, both of which were seeking to spread those standards more widely to other members of the EC.

Although operating under the rule of unanimity, the new authorization led to a series of EC-sponsored environmental programs in the ensuing years, which were aimed mainly at controlling water pollution, air pollution, toxic substances, and waste disposal.[31] Three action plans were adopted in the ten years following, giving rise to a stream of directives. It was apparent from the content of those programs that the EC had joined in the objective of controlling the degradation of the environment, though strongly preferring measures that did not inhibit cross-border movements within the common market. It was also apparent at times that the EC's involvement in environmental issues was having the effect of pulling up the standards of some member countries that might otherwise have been slower to act. One study, for instance, attributes Britain's conversion in the late 1980s from laggard to leader in a wide range of environmental issues in part to the pressures to which Prime Minister Thatcher was exposed in EC discussions.[32]

In 1987 the Treaty of Rome was amended by the Single European Act, which was aimed at providing the EC with added authority to override national regulations that inhibited the creation of a common market. For decisions aimed at achieving common market objectives, the principle of weighted majority voting was restored. Restrictions that inhibited the sale of services across national borders inside the common market were identified as a major target for EC action. Technical barriers imposed in the name of public health and consumer safety provided another major target.

At the same time, the Single European Act recognized environmental policy as a common policy of the EC, ostensibly equal in status to the creation of the common market. In the environmental field, however, the voting rules were complex and occasionally ambiguous: those actions that contributed to the common market objective could be taken by a weighted majority, but those that could not be justified in such terms still required a unanimous vote.[33]

Apart from ambiguities regarding the EC's voting rules in the exercise of its new responsibilities for the environment, other provisions of the Single European Act created additional uncertainties regarding the EC's writ in environmental matters. According to one such provision, the new powers of the EC were to be exercised with strict attention to the principle of subsidiarity; that is, they were to be applied only where an EC program could perform better than those pursued at the member state level.[34] Moreover, national governments were to retain the right to enact environmental provisions that were more stringent than those enacted by the EC, provided, however, that the provisions were consistent with the treaty. Finally, member states could enter into agreements with non-member governments and international organizations in the areas in which they retained competence.

It was obvious from the first that the Court of Justice would eventually have to take a hand in untangling the legal snarl created by these turgid provisions. Would the objective of achieving a single market prevail over the forces that seemed to be pushing in other directions, such as in the application of the subsidiarity principle and in the rights of member countries to apply their own strict environmental standards? By 1993 a number of cases that reflected the tensions among these different objectives were before the court for adjudication.

However, the enactment of the Single European Act was not the final episode in the evolution of the EC's powers over environmental issues. In December 1991, representatives of the twelve member states of the EU agreed to yet another round of treaty revisions (the so-called Maastricht treaty). One provision of the treaty deserves special attention. According to that provision, the principle of subsidiarity will apply not only in the field of environmental controls but in all fields in which the EU has competence.[35] That provision presumably adds to the weight and scope of the subsidiarity principle, placing added limits on the EU's ability to speak for Europe.

A CLOSER VIEW

To gain some sense of the decision-making process in the EU, one must constantly take into account the distinctive roles played by individual countries, such as the drive of Germany and the Netherlands toward higher environmental standards and the resistance of the poorer member states to such pressures. As discussed below, representatives of the member states permeate the decision-making processes of the EU. Without question, the hundreds of directives issued by the Council of Ministers under the powers bestowed by the Single European Act would not have survived without the support of the French president and the German chancellor.

The importance of the member states' role is particularly apparent in the enforcement of the EU's directives. Though the EU may command, it is for the member states to carry out those commands. In practice, national governments have commonly failed to meet their obligations, generating long delays and lawsuits brought by the commission before the Court of Justice. Indeed, in 1990 the commission had nearly four hundred cases outstanding that claimed noncompliance to its many environmental regulations and directives on the part of member states.[36]

Nevertheless, the institutions of the EU itself play a significant role, which needs to be distinguished from the direct and indirect actions of the member states. For instance, the programs developed under the Single European Act owed much of their scope and power to the ingenuity and entrepreneurship of a few key officials in the commission, including its chairman, Jacques Delors. Also, its initiatives could hardly have occurred without the ground-breaking decisions of the Court of Justice. Other EU institutions have also figured importantly in the shaping of the EU's role, in ways that have moved beyond the wishes of the member states. Even the EU's Council of Ministers, although ostensibly a body composed of national representatives, has produced results that were often more than simply a reconciliation of national positions.

The Council of Ministers, the law-making institution of the EU, is a body without precedent among international organizations. Though made up of ministers representing each of the member states, the identity of the ministers depends on the subject matter. Agricultural issues, for instance, are addressed by a council composed of agricultural ministers, environmental matters by environmental ministers, trade matters by trade ministers, and so forth.

One consequence of this odd structure has been to inhibit the give-and-take across functional areas that ordinarily occurs in policy-making bodies in democratic societies, such as national legislatures and national cabinets. While some member states, such as France, have created elaborate national institutions for coordinating their instructions to their Council of Ministers representatives, such national structures still cannot control the ultimate by-play on any issue that takes place inside the council. The fact that the deliberations of the Council of Ministers are held in secret has added to the sense that the laws they formulate are the product of a club made up of technocrats. There are indications that the secrecy surrounding the Council of Ministers' deliberations may lessen in the future, but the extent of such a shift is not yet clear.[37]

Although the Council of Ministers is the ultimate lawgiver in the EU, explicit proposals for the adoption of a directive must originate with the executive arm of the EU, a commission presently composed of twenty members nominated for fixed terms by the member states. The commissioners, who are assigned functional responsibilities inside the commission, are sworn neither to solicit nor to accept instructions from their governments. Although there have been issues on which some commissioners have been suspected of giving priority to national interests, by and large the commissioners have played independent roles as good Europeans.

Surrounding the commission and operating between that body and the council is the COREPER (Committee of Permanent Representatives to the Community), whose members are quintessential agents of the member states. Supported by a shifting network of subcommittees and working parties composed of national bureaucrats, the COREPER performs both an initiating and a lubricating function, advising the commission on initiatives the COREPER thinks would be welcomed by their respective governments, explaining the commission's positions to their respective ministers, and even acting for the Council of Ministers on a wide range of routine issues.

Finally, among the EU's key institutions is the European Parliament, an institution notable for the fact that it is the only body whose members are elected directly to the EU governing structure. The European Parliament is not a parliament of national parliaments. (Indeed, national parliaments are almost without power in the processes of the EU, except in the case of Denmark, whose ministers are bound to follow the instructions of their national parliament when engaged in the business of the EU.) The European Parliament is notable not only for its uniquely European credentials, but also because it is little more than an advisory body, empowered mainly to provide advisory opinions on the commission's proposals and on the council's prospective actions. True, it is empowered to take some substantive actions, such as to reject *in toto* the EU's proposed budget as well as to force the Council of Ministers to reconsider a decision if the commission also supports such reconsideration. Moreover, the Maastricht agreement slightly enlarges the parliament's powers, requiring the Council of Ministers to come to terms with the parliament if the council has not been unanimous in its decision. However, a unanimous decision of the council cannot be overridden by the parliament. The one body in the EU that the various national electorates might conceivably identify as truly European, therefore, has an inconsequential role in the decision-making process.

The unprecedented nature of the decision-making structure of the EU helps to explain its past role in matters of the environment. Its parliament, free to play the role of keeper of the European conscience, has been aggressive in its support of advanced positions on environmental controls.[38] However, it is unlikely that the parliament's aggressiveness would be quite so uninhibited if that body were to acquire substantial power to legislate. The EU's commission and its Council of Ministers have also been capable of substantial initiatives from time to time, as evidenced by the steady stream of environmental directives flowing from Brussels, and it could be that the heavy tilt in the EU's decision-making structure toward the functional approach has helped to give environmental ministers more freedom of action than they might have been able to exercise at home. But the many lapses of some of the EU's members in responding to its environmental directives and in enforcing those that have been enshrined in national law represent evidence of one of the structural limitations of the EU.

For the immediate future, two key factors have to be taken into account in judging the likely scope of the EU's authority to deal with environmental issues. One is the likelihood that the EU's relative immunity from the checks and balances that are an integral part of any democratic system may be drawing to an end, reducing the EU's ability to reach strong decisions on environmental matters. A second is the possibility that the Court of Justice's decisions on the enactment and enforcement of environmental provisions might alter the existing balance of power between the EU and its member states as well as among the various institutions of the EU.

ENVIRONMENTAL PROGRAMS

The cursory view that I have offered here of the decision-making processes of the governmental establishments in the United States, Japan, and Europe suggests that strong initiatives in support of international agreements on the environment may be forthcoming from time to time. In the United States, a necessary though not a sufficient condition for such an initiative might well be the decision of a high-level policy entrepreneur that the agreement was worth pursuing in order to make a personal mark. In Europe, an initiative from the EU might depend on the conclusion that an international agreement including outside countries would ease the problems of the EU in creating a single European market.[39]

Others who have studied the policy-making processes of the United States and other leading industrial countries feel that international initiatives on major environmental issues will be very slow to come. For instance, Eugene Skolnikoff, who has devoted a professional lifetime to the study of the formation of public policy in science, concludes that there will not be substantial international action on the global warming issue until the effects of the trend are far more palpable than they are today.[40] Skolnikoff's plausible views can be buttressed by another line of argument. For more than a decade, scholars have observed a decline in the public's confidence in the efficiency of the state in solving social and economic problems. That shift in public sentiment has served to arrest and even reverse the growth in the size and power of the public sector in the United States, Europe, and Japan, and it might reasonably be expected to stop the environmental movement in its tracks.

Yet, when the domestic environmental programs of the United States, Japan, and the EU are reviewed, they exhibit a vitality, a persistence, and a capacity for overcoming the resistance of groups with adverse economic interests that could not easily have been anticipated twenty years ago. Popular activism for the environment appears with a frequency encountered in very few other public issues, except perhaps that of abortion. A 1990 publication of the EU reports that the commission "receives an increasing number of complaints about the actual situation in the Member States from non-government organizations, local authorities, Members of the European Parliament, local pressure groups and private individuals."[41] In Europe, as well as in the United States and Japan, the political support that seems to provide a basis for environmental programs has often not come through the traditional party structures; it has mainly sprung up through organizations that have cut across political parties. Indeed, in Germany, Belgium, France, Ireland, Sweden, the Netherlands, and the United Kingdom, the environmental issue has led to the creation of new "green" parties that concentrate entirely on environmental issues. Although on the whole adherents to these green parties have leaned toward the left, their willingness to sacrifice growth for environmental betterment has generated great antagonism from some of the traditional left-leaning European parties.[42]

Can it be, therefore, that the analysis of decision-making processes that I have sketched out in the preceding pages will prove less important for environmental issues than for the usual business of governments? In projecting national behavior on environmental issues, perhaps one must give more than the usual weight to the possibility that groups outside the usual decision-making structures in the United States, Europe, and Japan will be able to overcome the resistance of those that see themselves bearing the costs of measures to protect the environment.

The U.S. role, as usual, is a wild card. The separation of powers in the U.S. system increases the difficulties of predicting its role. Observe, for instance, the refusal of the Bush administration to go along with definitive international commitments aimed at slowing global warming trends at a time when public sentiment appeared strongly to favor such commitments, or the initiative of U.S. representatives in securing international agreement to reduce the use of CFCs, notwithstanding the strong misgivings of industry regarding the effects of such restrictions. Predicting the U.S. role is complicated further by the independence of Congress and the courts in deciding just how to treat the country's international commitments in the face of domestic pressures.

As for Japan, there is nothing in the record to suggest that the decisionmakers who are normally in charge will show enthusiasm for international agreements on environmental measures. Initiatives are only likely for issues that are immediately related to the problems of living on the crowded islands of Japan. Moreover, the Japanese cannot be expected to embrace readily either specific standards or specific measures to achieve such standards, especially measures whose enforcement might involve adversary proceedings and adjudication. However, one cannot be sure whether a new generation of cosmopolitan Japanese will follow in the steps of earlier generations.

In the end, therefore, international activism on environmental issues may continue to depend on the out-of-channels pressures that appear to have been indispensable in the past. The scholarly study of that phenomenon is already well under way, breeding such ponderous concepts as the epistemic community.[43]

In a world in which the costs of international communication are plummeting and the facilities for communication proliferating, the growing importance of transnational groups with common views regarding the environment seems inevitable. Their appearance was heralded a few decades earlier by the proliferation of single-issue organizations in the political processes of various countries, such as the Sierra Club in the United States and the Greens in Germany, and followed by the development of strong links among these national organizations. The trend has been accelerated by the growth of boundary-straddling business organizations, such as the multinational enterprise. Such groups figured prominently in the treaty for the control of CFCs, as scientists and consumer groups exchanged information across borders and as DuPont's recognition of the CFC problem spurred British-based Imperial Chemical Industries to a shift in its position.

The idea that boundary-straddling communities may be critical in framing the content of international agreements, however, is an uncomfortable one. For one thing, this situation greatly complicates the structure of a well-known metaphor that political scientists have learned to apply when analyzing the international negotiating process—the concept of the two-level chess game. With these boundary-straddling communities included in the play, the international chess game begins to lose its adversarial character and takes on some of the qualities of a joint conspiracy against other players in national polities.

Boundary-straddling communities that represent relatively specialized interests have, of course, been common in the conduct of international relations in the past. Long before the identification of epistemic communities, the monthly meetings of the world's chief central bankers at Basle illustrated the importance of such institutions. In this case, the monthly meetings of central bankers evolved into an institution of some power and influence, which sometimes quietly engaged in a joint effort to bring free-spending national politicians to their senses. In addition, for many decades, the world's principal airlines—speaking in the name of their respective governments—jointly determined the prices that international travelers would have to pay for their services, justifying their decisions by a shared belief in the efficacy of the system. The influence of such groups may have served good purposes or ill, but obviously their views did not always reflect the diverse interests and complex priorities of the countries they purportedly represented.

In any case, although boundary-straddling communities may play a critical role in the adoption of new international agreements, they are likely to be much less effective in the implementation of such agreements. The adoption of new programs for the environment may be critical, but whether a regime can work is the ultimate test of its usefulness. In some environmental matters, for instance, the power to act or to avoid action is likely to revert to the day-to-day decision-making apparatus of the participating governments, including administrative agencies and the courts. In such cases, enforcement is likely to depend on mundane processes and incremental measures that do not easily arouse the enthusiasm and commitment of ordinary citizens. It is at this stage that understanding the history, institutions, and values of national decision-making institutions may prove especially critical.

ACKNOWLEDGMENTS

I have benefited enormously from perceptive reactions to earlier drafts of this chapter by Nazli Choucri, Henry Lee, Marc Levy, Kalypso Nicolaidis, Susan J. Pharr, Louis T. Wells, Jr., and Philip Zelikow.

NOTES

1. For an authoritative summary of such issues, see James K. Sebenius, "Designing Negotiations toward a New Regime," *International Security* 15, no. 4 (spring 1991): 110–148. See also Richard E. Benedick, *Ozone Diplomacy: New Directions in Safeguarding the Planet* (Washington, DC: World Wildlife Fund, 1990); Jessica T. Matthews, ed., *Preserving the Global Environment: The Challenge of Shared Leadership* (New York: Norton, 1990); and *Millennium* 19, no. 3 (winter 1990), a special issue devoted to global environmental change and international relations.

2. James W. Fesler, "Policymaking at the Top of Bureaucracy," in *Bureaucratic Power in National Policy Making,* ed. Francis E. Rourke (Boston: Little, Brown, 1986), 317, 330; and Roger B. Porter, *Presidential Decision Making: The Economic Policy Board* (Cambridge: Cambridge University Press, 1980), 5–21.

3. Hugh Heclo, *A Government of Strangers* (Washington, DC: Brookings Institution, 1977), 84–112.

4. Randall B. Ripley and Grace A. Franklin, *Congress, the Bureaucracy, and Public Policy* (Homewood, IL: Dorsey Press, 1980), 39, 45; and "Trading Places," *Washington Post* 14 December 1990, A25.

5. Stephen D. Cohen, *The Making of United States International Economic Policy*, 3d ed. (New York: Praeger, 1988), 39–41; and Raymond Vernon, Debora L. Spar, and Glenn Tobin, *Iron Triangles and Revolving Doors: Case Studies in U.S. Foreign Economic Policymaking* (New York: Praeger, 1991), 16.

6. The literature on this period is overwhelming in quantity and range. For a succinct summary of the international positions of the United States as they related to domestic politics, see Robert L. Paarlberg, "Ecodiplomacy: U.S. Environmental Policy Goes Abroad," in *Eagle in a New World*, eds. Kenneth Oye, Robert J. Lieber, and Donald Rothschild (New York: Harper-Collins, 1992), 209–231. See also Richard A. Harris and Sidney M. Milkis, *The Politics of Regulatory Change* (New York: Oxford University Press, 1989), 225–272.

7. See, for example, Thomas C. Schelling, *The Strategy of Conflict* (Cambridge: Harvard University Press, 1960), 27–28; Robert Putnam,

"The Logic of Two Level Games," *International Organization* 43 (1988): 439–440; and H. Richard Friman, "Rocks, Hard Places, and the New Protectionism: Textile Trade Policy Choices in the United States and Japan," *International Organization* 42 (1988): 689–709.

8. See, for example, James K. Sebenius, *Negotiating the Law of the Sea* (Cambridge: Harvard University Press, 1984), 71–109; and "Faltering: GATT and Services," *The Economist*, 14 July 1990, 70.

9. For a striking illustration, see Benedick, *Ozone Diplomacy*, 58–67.

10. Sheila Jasonoff, "American Exceptionalism and the Political Acknowledgment of Risk," in *Risk*, ed. Edward J. Burger, Jr. (Ann Arbor: University of Michigan Press, 1990), 63.

11. I. M. Destler, *American Trade Politics: System under Stress* (Washington, DC: Institute for International Economics, 1986), 83–86.

12. See G. John Ikenberry, "Manufacturing Consensus: The Institutionalization of American Private Interests in the Tokyo Trade Round," *Comparative Politics* 21, no. 3 (April 1989): 295–301; and Destler, *American Trade Politics*, 19–22.

13. The literature is extensive. See, for example, T. J. Pempel, "Japanese Foreign Economic Policy: The Domestic Bases for International Behavior," in *Between Power and Plenty: Foreign Economic Policies of Advanced Industrial States*, ed. Peter J. Katzenstein (Madison: University of Wisconsin Press, 1978), 139–190; Kent E. Calder, "Japanese Foreign Economic Policy Formation: Explaining the Reactive State," *World Politics* 40, no. 4 (July 1988): 517–541; Frank K. Upham, *Law and Social Change in Postwar Japan* (Cambridge: Harvard University Press, 1987); Daniel I. Okimoto, "Political Inclusivity: The Domestic Structure of Trade," in *The Political Economy of Japan*, vol. 2, eds. Takashi Inoguchi and Daniel Okimoto (Stanford, CA: Stanford University Press, 1988), 345–378; and in same volume, Donald C. Hellman, "Japanese Politics and Foreign Policy," 345–378.

14. See "And Japan's Quest for Autonomy," chapter 3 in Irvine H. Anderson, *The Standard-Vacuum Oil Company and the United States East Asian Policy, 1933–1941* (Princeton: Princeton University Press, 1975), 71–103.

15. Benedick, *Ozone Diplomacy*, 75.

16. See, for example, Fuchino Koichi, "Wage Earners' Changing Attitudes," *Japan Echo* 15 (1988): 17–23; and "Japan's Employers Come to Terms with Young Job Hoppers," *Financial Times*, 24 September 1991, 10.

17. Detailed descriptions appear in Shigeto Tsuru and Helmut Weidner, eds., *Environmental Policy in Japan* (Berlin: Ed Sigma Bohn, 1989);

Julian Gresser, Koichiro Fujikura, and Akio Morishima, *Environmental Law in Japan* (Cambridge: MIT Press, 1981); and Donald R. Kelley, Kenneth R. Stunkel, and Richard R. Wescott, *The Economic Superpowers and the Environment: The United States, the Soviet Union, and Japan* (San Francisco: Freeman, 1976). Especially useful is Susan J. Pharr and Joseph L. Badaracco, Jr., "Coping with Crisis: Environmental Regulation," in *America versus Japan*, ed. Thomas K. McCraw (Boston: Harvard Business School Press, 1986), 229–260.

18. See Margaret A. McKean, *Environmental Protest and Citizen Politics in Japan* (Berkeley and Los Angeles: University of California Press, 1981); and Norie Hudshle and Michael Reich, *Island of Dreams: Environmental Crisis in Japan* (Cambridge, MA: Schenkman Books, 1987), 14–23.

19. This interpretation draws heavily on Pharr and Badaracco, "Coping with Crisis."

20. For a detailed account of Japan's propensity to reduce conflict by payments to aggrieved parties, see Kent E. Calder, *Crisis and Compensation: Public Policy and Political Stability in Japan, 1949–1986* (Princeton: Princeton University Press, 1988).

21. See Edward B. Keehn, "Managing Interests in the Japanese Bureaucracy: Informality and Discretion," *Asian Survey* 30, no. 2 (November 1990): 1021–1037.

22. For the general tendency of Japan to maintain legal informality, avoiding hard-and-fast rules and standards, see especially Upham, *Law and Social Change*, 166–204. For these tendencies in the environmental field, see Shinichi Nakamura and Atsushi Toyonaga, "Making Environmental Policy in the United States and Japan: The Case of Global Warming," USJP occasional paper 91-08, Program on U.S.-Japan Relations, Harvard University, 1991.

23. For a comparison of U.S. and Japanese environmental movements, see McKean, *Environmental Protest*, 254–260.

24. Marc Levy generously gave me access to some early drafts of his uncompleted doctoral dissertation on environmental policies in Europe, which proved invaluable in bringing me up to date on some critical facts. He bears no responsibility, however, for my interpretation of these facts.

25. See Marc Levy, "The Greening of the United Kingdom: An Assessment of Competing Explanations" (paper presented at the 1991 annual meeting of the American Political Science Association, Washington, DC, 29 August–1 September 1991).

26. Carol Deck, "Negotiation and Compromise in German Environmental

Politics: Government, Industry, and Public," in *Germany at the Cross-roads: Foreign and Domestic Policy Issues*, eds. Gale A. Mattox and A. Bradley Shingleton (Boulder, CO: Westview Press, 1992), 149–161.

27. Compare Sheila Jasanoff, "American Exceptionalism and the Political Acknowledgment of Risk," *Daedalus* 119, no. 3 (fall 1990): 76.

28. See, for example, Arnold J. Heidenheimer, Hugh Heclo, and Carolyn Teich Adams, *Comparative Public Policy*, 3d ed. (New York: St. Martin's Press, 1990).

29. This tangled subject is well explored in Nigel Haigh, "The European Community and International Economic Policy," in *The International Politics of the Environment*, eds. Andrew Hurrell and Benedict Kingsbury (Oxford: Clarendon Press, 1992), 228–249; and Angela Liberatore, "Problems of Transnational Policymaking: Environmental Policy in the European Community," *European Journal of Political Research* 19 (1991): 281–305.

30. Neville March Hunnings and Joe MacDonald Hill, eds., *The Treaty of Rome Consolidated and the Treaty of Maastricht* (London: Sweet & Maxwell, 1992), Art. 36.

31. See Nigel Haigh, *Manual of Environmental Policy: The EC and Britain* (London: Longman, 1992). For a review, see J. Fairclough, "The Community's Environmental Policy," in *Britain, Europe, and the Environment*, ed. Richard Macrory (London: Imperial College, 1983), 19–34; David P. Hackett and Elizabeth E. Lewis, "European Economic Community Environmental Requirements," in "The European Economic Community's Product Liability Rules and Environmental Policy," Course Handbook No. 388 (New York: Practicing Law Institute, 1990); and Office for Official Publications of the European Communities, *Environmental Policy in the European Community*, 4th ed. (Brussels: 1990).

32. Levy, "The Greening of the United Kingdom," 20–28.

33. The revisions are found mainly in the Treaty of Rome Consolidated, Art. 130.

34. Ibid., Art. 130r.

35. Treaty on European Union (the Maastricht treaty) (Luxembourg: European Communities, 1992), Title II, Art. 3b, 13.

36. "The Dirty Dozen," *The Economist*, 20 July 1991, 52.

37. In a meeting at Edinburgh in December 1992, the European Council, composed of the heads of member governments, agreed to various measures that would slightly open the deliberations of the Council of Ministers.

38. See Peter Ludlow, ed., *The Annual Review of European Community Affairs* (London: Brassey's, 1991), xxx–xxxiii.

39. The EU's proposal to adopt a carbon tax provided that the United States and Japan follow suit probably represents just such a case. See "Europeans Still Agonizing over Carbon Taxes," *Financial Times*, 19 February 1993, 3.

40. Eugene B. Skolnikoff, "The Policy Gridlock on Global Warming," *Foreign Policy*, no. 79 (summer 1990): 77–93.

41. Office for Official Publications of the European Communities, *Environmental Policy in the European Community*, 31.

42. In a highly prescient article published in 1979, Suzanne Berger explored these trends in Europe (see "Politics and Antipolitics in Western Europe in the Seventies," *Daedalus* 108, no. 1 [winter 1979]: 27–50). Berger associates the trend with a general disillusionment with economic criteria as a basis for policy, thus offering an explanation of the survival and growth of environmental regulation.

43. For reasons that escape me, some scholars confine the concept of the epistemic community to organizations that challenge the "habit-driven behavior" of the existing decision-making apparatus; see Ernst B. Haas, *When Knowledge Is Power* (Berkeley and Los Angeles: University of California Press, 1990), 40–49. But these boundary-straddling communities may just as well be devoted to the preservation of existing habit-driven behavior, justifying their position on the basis of their common beliefs.

6

Trading in Greenhouse Permits: A Critical Examination of Design and Implementation Issues

■ **ROBERT W. HAHN**
The American Enterprise Institute

■ **ROBERT N. STAVINS**
John F. Kennedy School of Government
Harvard University

CONTROL OF GREENHOUSE GAS EMISSIONS

The desirability of taking immediate action to curb greenhouse gas emissions may be open to question, but it is important that the design and evaluation of policy options for addressing potential global change issues begin.1 Negotiators of future global warming treaties may examine a variety of international and domestic instruments to reduce greenhouse gas emissions. Among the approaches most likely to be considered are those that rely on command-and-control policies, which prescribe performance levels or technologies for specific sources, and those that use market-based instruments to provide economic incentives for achieving stated goals and standards. Two market-based instruments that are likely to receive great attention are tradeable property rights and taxes.

This chapter explores the possibility of nations using some form of tradeable property-rights approach to limit the level of greenhouse gases in the atmosphere. Although trading in greenhouse gas reduction permits has the potential to achieve a given level of reduction in emissions at least cost, there are many practical problems in design and implementation. This chapter frames these issues by examining the strengths and limitations of trading as a means of achieving a desired emissions limit for greenhouse gases both nationally and internationally. Thus, the central purpose of the chapter is to illuminate major design and implementation issues, not to advocate any particular type of permit system or even the use of such systems in general.

Tradeable permits could be used collectively by the community of nations, or individual countries could adopt permit-trading schemes to allocate control responsibilities domestically.[2] Trading could thus be used either on its own or in conjunction with some of the other policy mechanisms considered in this volume, including carbon charges. One issue that is critical for the success of any policy is the problem of how to allocate control responsibilities in a way that will allow negotiating parties to reach agreement. In this regard, we investigate how policies for trading in greenhouse gas permits can facilitate, or hinder, the likelihood that countries can reach agreement.[3]

In the next section, we propose a set of criteria by which alternative environmental policy mechanisms can be assessed. Then, following a brief review of conventional command-and-control regulatory approaches to environmental protection, we introduce the concept of incentive-based policy mechanisms and present the basic theory of tradeable permit systems. Next, we provide a review of actual experiences with these systems in the United States for air pollution control. We assess tradeable permits as a potential mechanism for addressing global climate change, describe some necessary conditions for successful permit markets, and highlight some crucial design issues that must be faced if a workable system is to be developed. Finally, we compare tradeable permit and charge mechanisms for the control of global climate change and offer some conclusions concerning the practical application of tradeable permits to the climate change issue.

To assess whether the trading of greenhouse gas reduction permits is likely to result in real advantages over alternative approaches, we consider the following criteria.[4]

- Will the policy achieve stated environmental goals?

- Will the policy approach be cost-effective? That is, will it meet environmental goals in the least costly manner?

- Will the strategy provide governmental agencies and private decisionmakers with information needed to implement the policy?

- Will monitoring and enforcement costs be reasonable?

- Will the policy be flexible in the face of changes in tastes, technology, resource use, and information?

- Will the policy give private industry incentives to develop new environment-saving technologies, or will it encourage firms to retain existing inefficient plants?

- Will the effects of the policy be equitably distributed?

- Will the purpose and nature of the policy be understandable to the general public?

- Will the policy be feasible? That is, can it be enacted, and can it be implemented by appropriate international agencies?

BASICS OF THE SYSTEM

As international deliberations consider various actions related to the threat of global climate change, policymakers will be faced with two tasks: the choice of the overall goal and the selection of a means, or instrument, to achieve that goal.[5] The two tasks are often linked in practice. For example, the goal might be to freeze carbon dioxide emissions levels at 1990 levels by the year 2000. The instrument used to achieve this goal could be tradeable permits, taxes, subsidies, conventional regulatory approaches, or some combination of these policies. Thus, market-based environmental policies are a means for achieving policy goals. Such policies are largely neutral with respect to the goal selected, though they can affect whether a particular goal is chosen.[6] Before investigating market incentives in general and tradeable permits in particular, we find it useful to review the regulatory approach most frequently used—that is, command-and-control regulations.[7]

CONVENTIONAL COMMAND-AND-CONTROL REGULATORY APPROACHES

A fundamental principle behind command-and-control regulations is that all firms should bear the same share of the pollution control burden, regardless of the relative costs of doing so. This regulation is typically implemented through the creation of uniform standards for firms, the two most prevalent of which are technology-based standards and performance standards. Technology-based standards identify particular equipment that must be used to comply with a regulation.[8] For example, electric utilities may be required to use electrostatic precipitators to remove particulates, or all firms in an industry must use the "best available technology" to control water pollution. Performance standards, on the other hand, set a uniform control target for each firm, while allowing some latitude in how the target is met. Such a standard, for instance, might set the maximum allowable units of pollutant per time period but be neutral with respect to the means by which a firm reaches this goal.

In contrast, by requiring firms to meet a specific standard, command-and-control regulations may result in unduly expensive means of a nation's achieving an environmental target.[9] The reason is simple: the costs of controlling pollutant emissions can vary greatly among and even within firms. Technology appropriate in one situation may not be appropriate in another. Indeed, the cost of controlling a given pollutant may vary by a factor of one hundred or more among sources, depending on the age and location of plants and the available technologies.[10]

A second shortcoming of this regulatory approach is that it tends to freeze the development of technologies that could provide greater levels of control. Because little or no financial incentives exist for firms to exceed their control targets, a bias exists against experimentation with new technologies (explicitly in the case of technology standards and implicitly in the case of performance standards). In fact, the reward to a firm for developing a new technology may be that it will subsequently be held to a higher standard of performance. Hence, dollars that could be invested in technology development are diverted to legal battles over what are or are not acceptable technologies and standards of performance.[11]

In terms of the performance criteria for good environmental policy presented above, command-and-control regulations can be designed to perform well on most of the criteria, except for cost-effectiveness and the incentives to develop new technology. Moreover, while it may be possible for command-and-control approaches to meet some stated environmental goals, the use of these regulatory approaches may slow the general rate of environmental progress.

INCENTIVE-BASED POLICIES

The primary motivation behind incentive-based policies is to save resources by encouraging more economical and effective use of scarce private and public funds. If it is possible to save money by lowering the costs of achieving an environmental goal, this action leaves extra funds that can be spent on other activities. Economists tend to support the use of these policies because they allow citizens and firms to satisfy more of their individual and social needs by enhancing the efficiency with which they use resources.

Incentive-based policy instruments save resources by equalizing the incremental amount that firms spend to reduce pollution—that is, their marginal costs of reducing pollution—as an alternative to equalizing the amount that firms pollute.[12] These instruments achieve the same aggregate level of control as a uniform standard, but with the control burden shared differently among firms. Firms that can control pollution at lower cost control their pollution more (emit less); those that have relatively high costs to control pollution control their pollution less (emit more). This situation results in a cost-effective outcome in which fewer economic resources in total are used to achieve the same level of pollution control.

Theoretically, the government could achieve such a cost-effective solution by setting different standards for each individual firm in a way that equated all firms' marginal costs of control. To do this, however, the government would need detailed information about the costs faced by each firm—information that the government clearly lacks and that it could obtain only at great cost, if at all. Fortunately, there is a way out of this impasse. Economic-incentive systems have the potential to produce the least costly allocation of achieving a given level of pollution control. By linking environmental performance with financial performance, these systems create incentives for firms to find cleaner production technologies.

While the precise mechanics of economic-incentive systems vary by type, they share a simple underlying principle. By making it costly for firms to pollute, the government makes it in the interests of firms and individuals to prevent pollution. The central role of government is to establish the incentives so that costs incurred by firms are sufficient to achieve the desired level of aggregate pollution control and (as in any regulatory system) to monitor compliance and enforce the law. Thus, economic-incentive systems do not represent a laissez-faire, free market approach. Instead, these systems recognize that market failures are typically at the core of pollution problems: the activities of firms and consumers have consequences for society (such as pollution) that are not adequately reflected in their own decision making. At the same time, incentive-based regulations reject the notion that such market failures justify scrapping the market system and dictating firm or consumer behavior. Instead, they provide freedom of choice to businesses and consumers to determine the best way to reduce pollution. By ensuring that environmental costs are factored into decision making, incentive-based regulations harness rather than impede market forces and channel them to achieve environmental goals at the lowest possible cost to society at large. In the language of economists, these systems internalize the externalities.

Most incentive-based approaches fall within five major categories: pollution charges, tradeable permits, deposit-refund systems, market-barrier reductions, and government-subsidy elimination. The first two categories—charges and tradeable permits—have received substantial attention as possible means of controlling global climate change.

With the charge approach, producers of pollution are charged a fee or tax on the amount of pollution they generate. This approach works on the assumption that producers will reduce pollution up to the point at which their marginal costs of control are equal to the pollution tax rate. As a result, firms will control to different degrees, with high-cost controllers controlling less and low-cost controllers controlling more. An effective charge system can thus minimize the aggregate costs of pollution control and give firms incentives to develop and adopt newer and better pollution control technologies.[13] However, pollution charge systems that are linked to emissions can impose significant monitoring burdens on governments.[14] Also, it is difficult to estimate in advance how large a charge will be required to obtain a desired level of pollution reduction, and it may be difficult, in a political context, to establish charges large enough to achieve given environmental objectives.[15]

In terms of the performance criteria for good environmental policy specified above, charges can be designed to perform well on most of the criteria, except for meeting specific environmental goals and, possibly, for monitoring and enforcement costs. In general, monitoring and enforcement costs may increase because firms can be expected to introduce a wide range of existing and new technologies to reduce pollution.

THE BASIC THEORY OF TRADEABLE PERMITS

Unlike a charge system, a system of tradeable permits allows the government to specify in advance an overall level of pollution that will be tolerated. This total quantity is allotted in the form of permits among polluting firms.[16] Under the most common type of system—that of emissions permits—firms that keep their emissions below the allotted levels may sell or lease their surplus allotments to other firms or use them to offset excess emissions in other parts of their own facilities. In this way, tradeable permit systems tend to minimize the total societal costs of achieving a given level of pollution control.

There are many points in the product cycle at which pollution can be regulated, whether with tradeable permit systems or otherwise. The simplest systems to organize and monitor (whether in the case of tradeable permits or other instruments) may focus on inputs to the production process, such as the lead content of gasoline or the carbon content of fossil fuels.[17] One step toward greater sophistication, but also substantially greater administrative complexity and transaction costs, is represented by emissions permits, which are tied to the quantity of pollutants actually emitted. Further in the same direction are ambient or concentration permits, which are tied to the effects that individual sources have on pollutant concentrations at specified receptor points. And further still are exposure permits and, finally, risk permits, the former referring to human or ecological exposures and the later referring to the consequences of such exposures.[18] Each successive system on this continuum may come closer to a theoretical ideal, but each system is also likely to bring greater public costs associated with monitoring and enforcement and greater private transaction costs. Indeed, these practical considerations help to explain why only input trading and simple emissions trading have actually been adopted or seriously considered by public authorities.

To see why tradeable permits are less costly than command-and-control approaches, consider the simple case of an industry with only two firms—one with a high marginal cost of control and the other with a low marginal cost of control. If permits are initially allocated uniformly to the two firms, it will be in the interest of the high-cost firm to purchase additional emissions permits and to increase its emissions as long as the price of the permits is less than the firm's marginal cost of pollution control. Likewise, it will be in the interest of the low-cost firm to sell additional emissions permits and to decrease its emissions as long as the price of the permits is more than the firm's marginal cost of pollution control. These competitive forces will tend to lead to an equilibrium permit price and distribution of permits (and pollution control responsibility), in which each firm equates its marginal cost of pollution control to the prevailing permit price. Therefore, the pollution control responsibility will be allocated in such a way that the two firms are controlling pollution at the same marginal cost of control (rather than at the same level of control). Hence the tradeable permit system, in theory, achieves the most cost-effective allocation of the pollution control burden among firms.[19]

In terms of the performance criteria for good environmental policy specified above, tradeable permits can be designed to perform well on most of the criteria, except, possibly, for monitoring and enforcement costs and accessibility to the general public.

U.S. EXPERIENCE WITH TRADEABLE PERMIT SYSTEMS

Tradeable permit mechanisms have previously been applied in the United States, including the EPA's Emissions Trading Program, the nationwide phasedown of lead in automotive fuel, CFC reduction, pointnonpoint source trading for water quality control, and, most notably, the 1990 amendments to the Clean Air Act, which include a permit trading system for sulfur dioxide emissions and the control of acid rain.[20] In the following section, we review four of these programs—the EPA's criteria for air pollutant trading, the phasedown of leaded gasoline, the trading in

CFCs, and the new sulfur dioxide trading system—to highlight lessons that can be learned about design and performance.

THE EPA'S EMISSIONS TRADING PROGRAM

In 1974 the EPA began to experiment with emissions trading as part of its efforts to improve local air quality. Under the various programs, firms that reduce emissions below the level required by law receive credits usable against higher emissions elsewhere. Under programs of "netting" and "bubbles," firms are permitted to trade emissions reductions among sources within the firm, so long as total, combined emissions comply with an aggregate limit.[21] Under the "offset" program begun in 1976, firms that wish to establish new sources of pollution in areas that are not in compliance with ambient standards are required to offset their new emissions by reducing existing emissions by an amount greater than the amount of pollution the new source will generate. This offset can be achieved through the firms' own sources or through agreements with other firms. Finally, under the "banking" program, firms may store earned emissions credits for future use, either to allow for internal expansion or for sale to other firms.

These programs were codified in the EPA's final policy statement on emissions trading in 1986, but their use to date has not been extensive.[22] States are not required to use the programs, and uncertainties about their future course have made firms reluctant to participate.[23] The programs appear to have been handicapped further by bureaucratic infighting and by opposition from environmental organizations.[24] Nevertheless, companies such as Armco, DuPont, USX, and 3M have traded emissions credits, and a market for transfers has arisen.[25] Even this limited degree of participation is estimated to have saved between $.5 and $12 billion since 1976.[26]

LEAD TRADING

The EPA's lead trading program contrasts with the emissions-trading program for other air pollutants and much more closely approximates the economist's ideal of a freely functioning market. The lead trading program was created to allow gasoline refiners greater flexibility during a period when the amount of lead in gasoline was being reduced to 10 percent of its previous level. Interrefinery trading of lead credits was authorized in 1982.[27] To create lead credits, refiners had to produce gasoline that had a lower lead content than required by the standard. Banking of lead credits (for later use) was initiated in 1985 and was used extensively by firms.[28] Unlike many other programs, the lead trading program was scheduled to have a fixed life from the outset. The trading program was terminated at the end of 1987, when the lead phasedown had been accomplished.

Although the benefits of the trading program are difficult to measure directly, it is clear that the program was successful in meeting its environmental targets.[29] The level of trading between firms was high, far surpassing levels observed in other environmental markets. In 1985 over half of the refineries participated in trading with other firms.[30] The EPA estimated that savings resulting from the lead trading program were approximately $200 million annually.[31]

TRADING IN CFCs

The EPA has implemented a market in tradeable permits to help comply with the Montreal Protocol, an international agreement aimed at slowing the rate of stratospheric ozone depletion. The Montreal Protocol calls for reductions in the use of CFCs and halons, the primary chemical groups thought to lead to ozone depletion.[32] The market places limitations on both the production and consumption of CFCs by issuing allowances that limit these activities. If a firm wishes to produce a CFC, it must have an allowance.[33]

The Montreal Protocol recognizes the fact that different types of CFCs are likely to have different effects on ozone depletion. The treaty assigns weights to each CFC on the basis of its depletion potential. The allowances required for a particular type of CFC are calculated on the basis of the impact this substance is thought to have on ozone depletion.

Through mid-1991 there have been thirty-four participants in the market and eighty trades.[34] However, the efficiency of the market is difficult to determine, in part because no studies were conducted that estimated cost savings or the likely pattern of trading. The problem is compounded because the timetable for the phaseout of CFCs has been accelerated and a tax on CFCs has been introduced. Indeed, the tax may now be the binding (effective) instrument.[35] Despite these confounding factors, relatively low transaction costs associated with trading in the CFC market may suggest a somewhat cost-effective outcome.

ACID RAIN CONTROL UNDER THE
CLEAN AIR ACT AMENDMENTS OF 1990

A centerpiece of the Clean Air Act Amendments of 1990 is a tradeable permit system to regulate sulfur dioxide, the primary precursor of acid rain.[36] Title IV of the act reduces sulfur dioxide and nitrous oxide emissions by 10 million tons and 2 millions tons, respectively, from 1980 levels.[37] The first phase of sulfur dioxide emissions reductions is to be achieved by 1995, with a second phase of reductions to be accomplished by the year 2000. In Phase I, 111 electrical utilities are targeted with individual emissions limits. After 1 January 1995, these plants can emit sulfur dioxide in excess of limitations only if they qualify for extensions or substitutions, or obtain allowances for their total emissions.[38]

Essentially, these allowances are permits to emit designated quantities of sulfur dioxide during or after a specified year. Under Phase I, the EPA will annually allocate to all affected plants a specified number of allowances related to its capacity, plus bonus allowances, which are available under a variety of special provisions.[39] Cost-effectiveness is promoted by permitting legal-allowance holders to transfer their permits among one another.

Under Phase II of the program, beginning 1 January 2000, almost all electric power generating units will be brought within the system.[40] Certain exceptions are intended to compensate for potential restrictions on growth and unfair treatment of units that are already unusually clean. If trading permits represent the carrot of the system, its stick is a penalty of $2,000 per ton of emissions exceeding any year's allowances (and a requirement that such excesses be offset the following year).[41] It is essential that noncompliance be unattractive if the system is to achieve its objectives.[42]

It is too soon to assess the program, but preliminary indications are that trading among utilities will be active, particularly in the second phase of the program when all utilities are brought into the allowance program, the reduction target is increased to 10 million tons, and trading on a futures market is implemented.[43] Among the major unanswered questions is how state regulatory commissions will allow utilities to treat the benefits and costs of sales and purchases of allowances. How these financial impacts are allocated between ratepayers and shareholders, for example, can have an important effect on the level of trading, as well as on the degree of cost-effectiveness that is achieved.[44]

LEARNING FROM EXPERIENCE

Marketable permits are rarely, if ever, introduced in their textbook form. Virtually all environmental regulatory systems using tradeable permits rely on the existing system of permits. This result should not be surprising, since these systems were not implemented in a vacuum. Rather, they were grafted onto regulatory systems in which permits and standards already played a dominant role.

Partly as a result of such hybrid approaches, the level of cost savings resulting from implementing marketable permits has varied widely but typically is far below their theoretical potential. The EPA's Emissions Trading Program has achieved only a small fraction of their potential cost savings. In contrast, the lead trading program enjoyed very high levels of trading activity, which suggests that the market was very efficient. The data on CFCs are more difficult to assess. Some trading has occurred, but it is unclear how much trading would be needed to reach the most cost-effective solution.

In most cases, marketable permit programs and command-and-control programs can affect environmental quality to roughly the same degree. In the case of lead trading and emissions trading, marketable permit programs have generally achieved similar environmental outcomes to the command-and-control programs they replaced. In the case of CFCs, there is no command-and-control benchmark for comparison; however, there is no reason to believe that the results of command-and-control regulations would have been significantly different from the market-based solution. Countries that agreed to phase out CFCs have generally complied with the spirit of the agreement, regardless of whether they chose to meet the phaseout using command-and-control or market-based approaches.

Although the impact of the acid rain allowance trading program on environmental quality is difficult to predict, it is unlikely that the targeted 10-million-ton reduction in sulfur oxides could have been achieved with a command-and-control approach. The political costs of an equivalent emissions reduction achieved through a command-and-control initiative of this magnitude would probably have been unacceptable. Either acid rain legislation would not have passed in 1990 or the emissions-reduction target would have been reduced.

In thinking about the environmental and economic characteristics of different regulatory approaches for environmental protection, it is important to take into account the political environment in which these approaches are implemented. Some of the design features of existing market-based approaches that affect economic performance are clearly dictated by politics.[45] An example is the bias toward scrubbing in the allowance trading market for the control of acid rain. At the same time, there is room within these political constraints for the design of more efficient markets. New environmental initiatives, which have not been subject to traditional forms of regulation, may afford greater opportunities for the implementation of market-based initiatives with fewer restrictions on trading dictated by political pressures. The control of greenhouse gases, of course, is an excellent example of a potential problem for which there is virtually no regulatory infrastructure in place.

TRADING IN GREENHOUSE PERMITS

A tradeable permit program for greenhouse gases would be a new application of the general approach that has been applied to lead permit trading among refiners, to transferable production permits for CFCs, and to sulfur dioxide trading for acid rain prevention. However, an international greenhouse permit trading program would not, by any means, entail a simple extension of the concept from these domestic programs to a global one. Concerns about proper design and implementation are well founded.[46] This section describes the fundamental properties of a system of international trading in greenhouse gas source and sink permits, reviews the potential attributes of such an approach, outlines some necessary conditions for a successful permit market, and highlights a number of major design issues.

AN OVERVIEW OF GREENHOUSE PERMIT TRADING

Economic-incentive policies in general, and emissions trading in particular, are well suited for the management of uniformly mixed air pollution problems, such as acid rain and global climate change, because these policy mechanisms allow for aggregate pollution reductions at minimum cost to society at large. With essentially uniformly mixed air pollutants, ultimate concern is on aggregate pollution levels, as opposed to specific emissions from individual sources.[47] It does not matter, for example, whether a reduction in carbon dioxide emission takes place in Thailand or the United States.

One potential application of tradeable permits in the greenhouse context is an international system of greenhouse gas emissions trading. An international tradeable permit system would essentially implement a regulatory program that sets overall limits on emissions of carbon dioxide or on combustion of fossil fuel or deforestation. Nations would be assigned a baseline that would establish the initial emissions levels from which reductions would be assessed.[48] Each country would have to meet its minimum standard, but could do so either by controlling emissions or by purchasing reduction credits from nations that exceeded their own standards.

After establishing responsibility among nations (as with the Montreal Protocol in the case of CFCs), the permits could be transferred among nations, thus ensuring that emissions-reduction goals are distributed in a cost-effective manner among participating nations.[49] Individual countries could thus achieve their respective control obligations through any means chosen. For example, the United States and other nations might choose to meet their control targets through some combination of carbon charges, domestic tradeable permits, or other incentive-based and conventional regulatory policies. In this way, an international carbon dioxide or greenhouse gas trading program could accommodate a variety of separate national implementation strategies. Such an international, government-to-government trading system raises a number of legitimate concerns, which we discuss later. Among these concerns are monitoring and enforcement problems and the likelihood that nations would not necessarily make cost-minimizing decisions about permit trading.

Tradeable permits could also be used as a domestic policy instrument to allocate the control burden among firms and individuals in the pursuit of a national goal for reducing greenhouse gases (presumably set by international agreement). In this case, the most likely trading system would not focus on emissions, but rather on the carbon content of fossil fuels. In a way analogous to the EPA's leaded gasoline phasedown, the trading currency would be units of carbon in coal, petroleum, and natural gas, whether these fuels are domestically produced or imported. Such a system is closely related to the charge system most frequently considered for greenhouse gas reduction, namely, a carbon tax. As in that case, it is conceivable that units of carbon bound for manufacturing use (that is, nonfuel uses) would be exempted from the system.

Midway between international (government-to-government) trading and domestic (firm-to-firm) trading is a spectrum of mixed approaches that would allow firms or individuals in one nation to trade with firms or individuals in other nations. Although this increased flexibility is desirable from some perspectives, it raises some special problems if it is carried out in the context of other active greenhouse policy instruments, such as domestic carbon or energy taxes.[50] Because of this, we focus on two prototypes—international carbon dioxide emissions trading and domestic carbon rights trading.

Depending on the type of system adopted, there are a number of potential advantages in the use of marketable permits to reduce greenhouse gas emissions.

- *Flexibility.* Businesses and individuals could have greater flexibility in choosing tactics and strategies for limiting greenhouse gas emissions. This flexibility could allow governments to focus on monitoring and enforcement rather than on writing detailed regulations that prescribe specific technologies for energy efficiency or pollution reductions.

- *Direct control of aggregate emissions levels.* Essentially, marketable permits place a cap on emissions levels. The number of permits issued determines the aggregate level, provided these permits are effectively enforced.

- *Cost-effectiveness in pollution control.* Markets provide positive incentives in the form of cost savings and profit for individuals and businesses to seek out the lowest cost methods of achieving emissions targets (if the trading parties are themselves cost minimizers).

- *Provision of mechanisms for trading among different greenhouse gases.* Allowing trading among different gases would enhance opportunities for cost savings. A key issue is whether such trades can be justified on scientific grounds.

- *Dynamic incentives for the development of low-polluting technologies and management strategies.* A major advantage of market-based approaches is that they provide incentives for the development of innovative approaches to emissions reductions that are largely absent from command-and-control regimes.

- *Establishment of a mechanism for directly addressing the equity concerns of developed and developing countries.* Policymakers can distribute permits in a way that promotes fairness, however they choose to define it.

- *Linkage of sources and sinks in a single, comprehensive strategy.* Theoretically, it would be desirable to link sources and sinks so that overall emissions limits could be achieved at the least cost. Some sinks, such as forests, pose formidable obstacles in measurement and enforcement.

Even a limited subset of these potential advantages of tradeable permit systems can be realized in practice only if very careful attention is given to a host of serious design and implementation issues. It is to such issues that we turn next.

CONDITIONS FOR A SUCCESSFUL PERMIT MARKET

The full measure of potential cost savings offered by tradeable permits in greenhouse gases can be realized only if efficient markets develop. Four conditions are necessary for this to occur: (1) a sufficient degree of compliance must be achieved, (2) transaction costs must be low enough not to prevent efficient permit exchanges from taking place, (3) the market for permits must be competitive, and (4) the policy must be seen as one that will remain in place for a significant period of time.[51]

MONITORING AND ENFORCEMENT: THE QUESTION OF COMPLIANCE

For a tradeable permit (or any other) policy to be truly effective and to operate at least cost, firms or nations must comply with the policy's requirements. In the international trading context, this requirement means that nations must accurately report changes in greenhouse gas sources and sinks and must purchase the appropriate number of permits if their net emissions exceed their permitted level. The costs for an international body to monitor and enforce this system would be significant, but such compliance problems would arise with a number of alternative approaches as well. In the domestic-trading context, monitoring and enforcement problems and related costs are inextricably linked to the level of trading chosen. Trading in carbon-content rights for the three fossil fuels among producers and importers would likely involve much lower monitoring and enforcement costs than would emissions trading.

TRANSACTION COSTS

The potential cost savings associated with tradeable permit systems can be realized in practice only if trading involves sufficiently low transaction costs, including the costs of finding prospective buyers and sellers as well as the costs of obtaining any necessary approval for trades.[52] If nations or firms lack information on other countries or firms, identification would be costly; but in the presence of potentially high transaction costs, brokers would likely emerge to facilitate trading by linking potential buyers and sellers of permits.[53]

Regulatory constraints could lead to high transaction costs in an international tradeable permit market if governments were required to obtain approval from an international body for each and every transaction. Such regulatory constraints would likely be most significant where localized environmental impacts are important in order to ensure that hot spots of emissions do not occur; however, given the global commons nature of the impacts of greenhouse gases, these hot-spot effects are not an issue.[54]

Since transaction costs are partly a function of the number of potential traders in a relevant market, these costs (like those associated with monitoring and enforcement) will be closely linked to the chosen level of trading activity. In the case of domestic trading, this clearly argues in favor of carbon-input trading, as opposed to carbon dioxide emissions trading.[55]

COMPETITIVE MARKET CONDITIONS FOR PERMIT EXCHANGES

The degree of competition in the permit market will also affect the extent to which potential cost savings are likely to be realized.[56] Under perfectly competitive conditions, each nation or firm would decide whether to enter the permit market depending on the market price of permits versus its internal costs of controlling sources and sinks of greenhouse gases. However, a nation or firm that buys or sells a significant fraction of the total number of permits traded might be able to influence their price. If prices were manipulated by a particular firm or nation, then greenhouse gas reductions would not be achieved at minimum cost.[57]

CERTAINTY IN THE PERMIT MARKET
AND THE WILLINGNESS TO TRADE

Firms and nations cannot be expected to engage in trading if they believe that the policy is likely to change dramatically. Consequently, policies must be stable for a reasonable period of time. However, governments may have a great deal of difficulty in creating an international agreement that private parties believe will remain unchanged, which could have a significant adverse impact on parties' willingness to engage in cost-effective investments in reducing greenhouse gas emissions.

Even if a policy is viewed as stable by key participants, firms and nations can be expected to trade greenhouse gas permits only if the rights these permits bestow are clearly defined and only if there is little uncertainty regarding the legitimacy of transactions. One potential source of such uncertainty for permit buyers in the international-trading context is whether nations that sell permits have actually made the necessary net reductions in greenhouse gases. A possible solution is to place responsibility for permit legitimacy with sellers rather than with buyers. Permits could be registered with an appropriate international body with sole enforcement responsibility. This scenario returns us, however, to the question of enforcement. Since it is by no means clear what international body could credibly monitor and enforce a tradeable permit system (or any alternative approach), the true practicality of this and other instruments cannot be fully assessed at present.

DESIGNING A GREENHOUSE TRADING SYSTEM

For the development of an effective and practical trading system, consideration should be given to a number of particularly important issues regarding policy design. Among these issues are (1) who should decide whether trading is allowed, (2) who should participate in trading, (3) at what level trading should take place, (4) whether there should be exclusive source programs or source and sink programs, (5) whether there should be carbon dioxide or multiple-gas trading, (6) what the aggregate target levels and initial allocation of control responsibility among nations should be, (7) how technology transfers to developing countries should be handled, (8) what approaches to issuing permits and to monitoring and enforcement should exist, (9) what temporal issues and permit banking possibilities should be considered, and (10) how alternative policy instruments should be adapted to changes in scientific understanding.

THE DECISION ON WHETHER TRADING IS ALLOWED

The decision to select a particular domestic regulatory approach, such as trading, should be left up to each country. Thus, an international agreement to limit greenhouse gases should allow nations to implement trading as a means of achieving their objectives. Moreover, if a country signs such an agreement, it should be responsible for demonstrating that it will meet the terms of the agreement using approaches permitted by the agreement. An international trading system should also be voluntary. Nations need not participate unless they so choose (as long as they comply with internationally mandated targets).

PARTICIPANTS IN TRADING

Negotiators will have to decide who should participate in trading. With regard to domestic trading, any individual, business, or public entity should be allowed to participate in programs, subject to constraints related to administrative feasibility. For example, obviously, it would be impractical to make individuals accountable for their carbon dioxide emissions. The question of how to draw the line to make the regulatory scheme manageable is an important one. Some have advocated starting off with a simple scheme, even if it does not encompass a large part of the problem.[58]

The problem of international trading is more subtle.[59] If international trading takes place among nations, there would be advantages of allowing firms and individuals within those nations to trade internationally. But depending on the nature of domestic greenhouse regulations, this allowance could raise serious problems of departures from cost-minimizing allocations of the control burden. In any event, limitations on trading may be necessary because of the potential for high administrative costs. On the other hand, it can be argued that national governments ought not be the primary agents for trading because they may not have the appropriate incentives or information needed to minimize costs. A large part of the decision of whether to allow various kinds of international trading will hinge on the differential ability and willingness of nations to enforce this kind of an agreement, a point that is addressed below.

Some have questioned whether developing countries should be permitted to participate in international trading because of fears that these countries, or individuals within these countries, might trade marketable permits to others to raise cash quickly without having taken adequate stock of long-term consequences. However, this problem may arise not because of a failure in these nations to take a long-term view, but rather because of the difficulty in these countries of implementing effective enforcement strategies, particularly in nations lacking stable political structures. The challenge is to develop international enforcement mechanisms that are credible, so that countries and individuals understand that not playing by the rules will entail real costs.

THE LEVEL OF TRADING

As mentioned earlier, there are a variety of points in the product cycle at which tradeable permits can be assigned to regulated parties. A system that merits consideration in the domestic context is a system of trading carbon rights in fossil fuels in which the rights are required both for the production of primary fuels at the mine mouth or wellhead and for importation. Such an approach would be similar to the lead rights trading program of the 1980s and is parallel to the type of charge system most frequently considered for carbon dioxide control, a carbon tax. If the permits were freely distributed, then the permit system would resemble a carbon tax system in which revenues were redistributed according to initial production and import levels.[60]

On the other hand, a system of international trading among governments of "control responsibilities" or targets could conceivably be linked to emissions, although the monitoring and enforcement problems are likely to be massive. In any event, it is safe to say that most attention ought to be focused on input and emissions trading, since in the case of this uniformly mixed global pollutant, ambient trading would be an unnecessary and costly complexity. The same can certainly be said of exposure and risk trading.

SINKS AND SOURCES

Should an international carbon dioxide trading program only consider changes in emissions of carbon dioxide or also consider changes in carbon dioxide sinks (such as expanding forests)? Since growing forests remove carbon from the atmosphere and their burning or destruction contributes to global carbon dioxide loadings, one question is whether and how an international agreement might help retard deforestation and promote reforestation. With a tradeable permit program for carbon dioxide, countries like Brazil and Indonesia might find it economically attractive, as well as environmentally sound, to retard the depletion of their forests or to implement reforestation programs in order to earn carbon dioxide credits, which their own industries could use or which they could sell to foreign governments.[61] Those countries in turn could use the revenues from the sale of such credits to finance programs to retard forest loss.[62] Depending on the initial distribution of carbon dioxide permits, which determine reduction responsibilities, an international trading program could contain an explicit mechanism for addressing issues of equity raised by developing countries.[63]

The inclusion of sinks, however, would entail many complications. First and foremost is the issue of the need to establish a relevant baseline. How many trees are out there now, and who should get the credits for these trees? Second, there is the issue of how to measure changes in greenhouse gas emissions associated with various sinks. It is not a simple matter to measure the carbon sequestered in trees and how it varies over time.[64] Moreover, should individuals get credit for planting trees in their backyards? Suffice it to note that sinks frequently pose formidable measurement problems and are often owned or managed by governments. While sinks can be included in a market-based approach in principle, caution is required before they are included on a large scale, at least in the short term.

CARBON DIOXIDE VERSUS MULTIPLE GREENHOUSE GAS TRADING

There has been considerable debate in the United States regarding whether to focus on carbon dioxide alone for a trading program or whether to include all greenhouse gases.[65] If there were trading in all greenhouse gases, should there be trading across gases as well? Trading across gases adds an additional layer of complexity because it requires the establishment of trading ratios.[66]

The advantage of the more comprehensive approach is the additional flexibility it introduces into the system, and hence the potential it creates for even greater cost-effectiveness.[67] On the other hand, such a system would increase administrative burdens and require greater scientific understanding of the relative impacts of the suite of greenhouse gases. One option is gradually to introduce other gases into the trading framework, contingent upon increased scientific knowledge about these gases.[68] Experience suggests that when there is significant profit potential and when it is clear that regulatory action will be taken, firms or nations may cooperate in the development of baselines, budgets, and appropriate certification methods.[69]

Given the current understanding of the causes and consequences of global climate change, and the technological limitations, a comprehensive market-based approach for greenhouse gas emissions is probably not appropriate, even in the United States. This is not to suggest that all gases should not be considered in an agreement, but that there are significant limitations to markets when monitoring and enforcement are particularly costly and when the science is highly uncertain.

AGGREGATE TARGET LEVELS AND INITIAL ALLOCATION OF CONTROL RESPONSIBILITY AMONG NATIONS

Any international greenhouse gas reduction policy presumes some agreed-upon aggregate goal for the management of greenhouse gas emissions into the atmosphere, which is translated into an allowable level of emissions over time. Initially, the goal can be framed as a fixed level of ambient greenhouse gas concentrations.[70] This level can be established as a reduction from a baseline, either historic or projected. Once established globally, however, these allowable emissions must be distributed to individual nations.

The most difficult problem associated with any international greenhouse gas control program will be negotiators achieving agreement on both the global emissions cap and the initial control obligations of individual nations. The trading program highlights this problem because it makes it explicit. Since the program would create a new environmental currency denominated in tons of carbon dioxide (or other gases), every nation will immediately know its reduction responsibilities.

A variety of alternative allocation mechanisms have been suggested, including allocations based on GNP, real GNP, total population, adult population, land area, and emissions.[71] There are numerous other possibilities. Each of these criteria will have adherents, largely those with larger allocations under that criterion.[72] Several criteria may need to be blended to create international consensus on emissions allocations.[73] For example, developing countries will have relatively little incentive to participate unless they see clear economic benefits from an agreement. At the same time, wealthy countries will want to ensure that their burdens are divided in ways that are perceived as equitable. Whatever the initial allocation, subsequent trading can lead to a cost-effective outcome.[74] This potential for pursuing distributional objectives while assuring cost-effectiveness is an important attribute of the tradeable permit approach.[75]

TECHNOLOGY TRANSFERS TO DEVELOPING COUNTRIES

A trading system could provide industry with economic incentives to develop and use more efficient energy technologies and to switch to non-fossil or less carbon-intensive fuels.[76] Internationally, an appropriately designed trading program could promote the transfer of energy-efficient technologies from highly industrialized to developing countries.[77] For example, the potential of a developed country to obtain credits in a developing country by investing in increased appliance efficiency could create an economic incentive on the part of firms in industrialized countries to transfer technology and, in effect, finance the transfer of that technology. A well-designed protocol that encourages such international trading of energy credits could promote least-cost energy efficiency investments (as well as renewable energy investments) on an international basis. As a general proposition, the tougher the carbon dioxide reduction goals that industrialized countries must meet, the more they will be inclined to look for opportunities in developing countries as a source of credits.

Monitoring and enforcement requirements

The nature and magnitude of monitoring and enforcement costs associated with a greenhouse gas tradeable permit system will depend on the level of trading (for example, emissions permits versus carbon rights) and on the character of participants in the program (that is, nations, firms, or individuals). In the domestic-trading context, there is little doubt that the monitoring and enforcement costs of a carbon-rights trading program would be significantly less than would be those associated with some sort of carbon dioxide emissions trading system.[78] Indeed, the overall administrative costs of a carbon-rights trading program should be no more than those of a system of carbon taxes.

In the context of international emissions permits, the authority to issue permits would presumably come from an international agreement on climate change. Some existing or new supranational body could, theoretically, be vested with the authority to issue permits. Nations could then develop regulatory approaches of their choosing to meet the limitations on greenhouse gases that are imposed by the permits. For such a system, even basic monitoring of compliance will be a formidable challenge. One only need take note of the widely varying estimates of Brazil's rate of deforestation to appreciate the magnitude of the problem.[79]

Clearly, there are significant trade-offs between monitoring ease and accuracy. For example, it would be theoretically desirable to allow full flexibility for nations to achieve their emissions targets (permit levels) through any means they might choose, including reduced fossil fuel combustion on the one hand and reforestation on the other. But this process would also necessitate an extremely expensive monitoring system *or* the adoption of some simplifying assumptions (regarding, given the above examples, the impact of given fuel uses on emissions and the relative impacts of various reforestation programs).

Two critical problems are likely to emerge in enforcement within the international emissions-trading context. First, enforcement will differ significantly across countries that sign an agreement. Second, countries that do not sign an agreement to reduce greenhouse gases will be able to free ride on those that do. Neither of these problems has simple solutions that are politically feasible in the near term.

One possible approach linking enforcement to international trading would be to restrict international trading to those nations that are deemed to be in compliance by some supranational group.[80] For example, if the supranational group found that Germany and the United States were both in compliance with their obligations, then trading would be allowed between the two nations. A problem with this approach could arise if countries change their level of effort in enforcing an agreement. If, for example, a country traded away permits while in compliance, but then subsequently went out of compliance with an international agreement, rules would need to be developed to define which country would be responsible for limiting its greenhouse gases to come back into compliance. While this problem is manageable in theory, it could quickly lead to a political quagmire. For example, the supranational authority could be asked to make difficult decisions about whether to allow developing countries, with limited resources for monitoring and enforcement, to engage in international trading.[81]

A somewhat more ambitious alternative would allow international trading based on allowances that do not trade on a one-for-one basis but that reflect their actual value (rather than the face value of the allowance). Thus, if Denmark were in compliance, but Brazil had greenhouse gas emissions that were 100 percent above its allowable levels, then Brazilian permits would be traded at a 2:1 ratio with Danish permits. Again, the supranational group would need to define the trading ratios between nations. This latter approach adds a level of complexity that may make the system very difficult to administer. Moreover, countries would probably be reluctant to cede this kind of authority, with its potential for large implicit wealth transfers, to an international body.

As technology improves for monitoring carbon dioxide and other greenhouse gases, it may be easier to reach agreements that would allow for international trading with modest transactions costs. At present, both technological constraints and political forces pose formidable obstacles to the emergence of a workable international emissions-trading regime.

TEMPORAL CONSIDERATIONS AND THE POSSIBILITIES OF PERMIT BANKING

The temporal component of any pollution problem can be important but is particularly so in the case of stock pollutants, which tend to accumulate in the environment at a rate that significantly exceeds their natural rate of decay. Accumulations of greenhouse gases are of this nature and thus raise a set of time-related issues. If the overall goal of some public policy were to limit the rate or degree of climate change, significant trade-offs would exist with regard to the timing of any proposed reductions in greenhouse gas emissions. Earlier reductions would have the effect of slowing the potential onset of climate change.

Within the context of a tradeable permit system, these temporal considerations can be addressed, to some degree, through provisions for (or restrictions on) banking, a mechanism that enables firms or nations to make early emissions reductions in exchange for the right to emit a comparable amount at some later date. This notion could be extended to sinks as well as sources. It could be advantageous to allow nations to engage in the banking of greenhouse gas allowances, since this practice would allow for intertemporally efficient market exchanges and would tend to delay the onset of global climate change.

On the other hand, the ability to hold permits over time raises concerns about possible strategic behavior in which nations or sources that receive large initial allocations may choose to hoard their permits in order to exercise market power. Regulated leasing of permits could be used to address these concerns, at least in principle.[82]

THE ADAPTABILITY OF POLICY INSTRUMENTS TO EVOLVING SCIENTIFIC UNDERSTANDING

A critical challenge in the design of greenhouse regulatory instruments is to make sure that they are capable of responding appropriately to the evolving scientific and economic understanding of global climate change and its causes and consequences. Marketable permits are somewhat flexible in this regard. The authority that issues permits can make clear the conditions under which they will remain valid. But placing too many contingencies on permits is dangerous. Firms may not find it in their interest to participate in such a market. Governments have a reputation for not being able to make credible commitments honoring the value inherent in tradeable permits. Indeed, in some cases, they have purposely discriminated against permits that have been traded.[83]

Nations may want to consider two strategies. The first is to use markets for greenhouse gases where the science is reasonably certain and the problem can be measured with a reasonable degree of precision. This strategy suggests that carbon dioxide may be the most promising candidate for the application of some kind of incentive-based approach.[84] A second strategy that nations could pursue is to agree to revisit the issue at fixed intervals in order to modify strategies with evolving scientific and economic understanding. This approach would provide firms and nations with some guarantee about the value of their allowances over a specified time. At the same time, it would provide nations with needed flexibility to adapt to the changing science.

Whether these and other design problems will overwhelm the potential advantages offered by greenhouse permit trading programs will be determined by the specifics of alternative instruments under consideration and by the skills of those negotiating the framework agreements.

THE CHOICES

Many of the current discussions regarding alternative mechanisms to address global climate change have focused on the potential use of carbon taxes.[85] Therefore, in this final part of the chapter, we compare charges and tradeable permits as practical mechanisms for addressing global climate change, examine some of their political implications, and offer some brief conclusions.

COMPARING TRADEABLE PERMITS WITH CHARGE SYSTEMS

In many situations, economic-incentive policies can lead to lower total pollution control costs and can spur greater technological innovation than can conventional command-and-control approaches. However, which incentive-based instrument is most appropriate depends on a number of specific factors. We consider the broadest set of possible international and domestic applications.

1. *Both charges and permits impose costs on industry and consumers, but the costs associated with charges are more visible.*[86] Both charges and permits force firms to internalize the costs of their pollution. Practically speaking, this internalization of costs means that firms will experience financial outlays, either through expenditures on pollution controls or through cash payments (buying permits or paying charges). Charge systems tend to make these costs more visible to industry and to the public. This visibility may be problematic for political reasons, although in the long run it has the advantage of clearly signaling and educating the public about the costs and trade-offs associated with various levels of environmental control.[87]

2. *With permits, resource transfers are between private parties, whereas they are*

typically from a private party to the government with pollution charges.[88] Under permit trading, firms choosing to emit pollution beyond their initial permitted level must make payments to other firms that agree to control more than their initial share. With charges, payments for uncontrolled emissions flow to government. For those who believe that the private sector can utilize the resources more effectively, permits offer an advantage over changes. On the other hand, revenue collected from charges can be earmarked for environmental investments, deficit reduction, or reductions in distortionary taxes.

3. *In theory, permits fix the level of control, whereas charges fix the marginal costs of control.* Under a permit system, policymakers determine how much total pollution can occur (through the issuance of permits), but they do not and cannot set bounds on expenditures on pollution control. Such a strategy could be particularly appropriate for environmental problems that exhibit tight margins of error regarding emissions levels or in which the marginal costs of control do not increase dramatically with increasing regulatory stringency.[89] Charges, on the other hand, control the maximum amount that a firm may pay for each increment of emissions, but they do not dictate with certainty how much control will actually occur. Such a tactic may be more appropriate where the margin of error on damages is not tight but where the potential impacts on industry of overcontrol are especially great.[90] This result could occur, for example, when small increases in control costs lead to large swings in production and employment. Which case is closer to that of global climate control will be clarified only through future research.

4. *Short of additional governmental intervention, permits freeze the level of control while charges increase it over time in the presence of technological change.* With a permit system, technological improvement will normally result in lower control costs and falling permit prices rather than in declining emissions levels. Such technological change under a charge system, however, will lead to both lower total control costs and increased levels of control. As technological change pushes the costs of controlling emissions down, firms will choose to control more emissions and pay less taxes.[91]

5. *Permits adjust automatically for inflation, whereas charges do not.* Because the currency under a permit system is emissions rights, levels of emissions control are unaffected by price movements in the overall economy. This is not the case for pollution charges. General price inflation will have the effect of reducing taxes (which are expressed in

dollars per ton, for example) in real terms. Therefore, in an inflationary environment, firms will control less than if prices were constant.

6. *Permit systems may be more susceptible to strategic behavior.* In order for a permit system to work effectively, relatively competitive conditions must exist in the permit (and product) market. The degree of competition will help determine the amount of trading that occurs and the cost savings that will be realized. Should any one firm or nation control a significant share of the total number of permits, its activities may influence permit prices.[92] National governments or firms might attempt to manipulate permit prices to increase their positions in the permit market or in the final product market (for example, by withholding permits and forcing others to cut production or by keeping new entrants out).

7. *Transaction costs can be important; they depend, in part, on the number of traders in the market.* Transaction costs (such as the costs associated with the identification of willing buyers or sellers of permits, or the costs of tax collection) not only drive up the total costs of compliance of incentive-based mechanisms, but also affect the amount of trading that will occur in a permit system and the amount of pollution control that will be achieved with a charge system. Hence, the relative magnitude of transaction costs is another important determinant in the identification of the preferred mechanism. In most cases, a comparison of transaction costs will tend to favor a charge approach.[93]

THE POLITICAL ECONOMY OF CHARGES AND PERMITS

All economic instruments, including charges and tradeable permits, are considered by policymakers in an intensely political environment. This reality has several important implications for the nature of these instruments, as well as for the potential for reaching an international agreement on climate change.

1. *Which policy instruments are used to achieve goals can themselves affect the likelihood that an agreement will be reached.* As mentioned earlier, in the domestic context it is unlikely that a 10-million-ton sulfur dioxide reduction goal would have been adopted in the United States had tradeable permits (and the cost-effectiveness associated with them) not been part of the legislative package. In the international context, it should be noted that under certain conditions a tradeable permit system can have the effect of reducing the min-

imum number of countries necessary for a global agreement to be reached.[94]

Countries can affect negotiations by the stand they take on the application of regulatory instruments to achieve environmental goals. If, for example, the United States continues to insist that all signatories to an agreement limiting greenhouse gas emissions must use market-based approaches for achieving the desired targets, this approach could either reduce the probability of achieving consensus or reduce the number of countries willing to sign an agreement.[95]

2. *The enforcement of instruments is likely to vary dramatically across nations.* This statement is true for both charges and marketable permits, as well as for command-and-control approaches. Its truth results from the fact that nations are unlikely to grant significant authority to a supranational unit that would allow for consistent enforcement across countries.

3. *Any market-based approach that is implemented to control greenhouse gases, even within a given country, will vary dramatically from the textbook application of these concepts.* There are many reasons why market-based approaches will deviate from their ideal; an important one is politics. Past examples include special preferences for certain technologies in acid rain trading and special exemptions for taxes on CFCs. However, departure of actual instruments from a theoretical ideal is not just cause, on its own, for rejection of the approach.

4. *Any market-based approach that is implemented will have to acknowledge implicit claims on wealth associated with the current distribution of greenhouse gases.* Pollution control strategies that have used charges and permits share the feature that at least some of the wealth that inheres in these approaches remains with the affected parties. Thus, for example, many charge systems in Europe designed to limit pollution recycle revenues to the participants or earmark the revenue for specific tasks. Similarly, in the United States tradeable permits for protecting the environment are distributed on the basis of the historical pattern of emissions. While the precise nature of the distribution will be the subject of vigorous political discussions, countries and special interest groups (including environmental groups) will not accept an agreement that substantially shifts the distribution of wealth or political power. This resistance means that market-based approaches may be more controversial than command-and-control approaches, but it also means that market-based approaches can facilitate the formation of coalitions of support through the grandfathering of rights.

5. *A market-based approach for limiting greenhouse gas emissions is likely to have more stringent monitoring and enforcement requirements than would a command-and-control system.* For example, environmentalists bargained for the installation of continuous emissions monitors as a condition for allowing a tradeable allowance system for the reduction of sulfur dioxide emissions in the United States. A similar strategy is likely to be applied if market-based approaches are implemented for limiting greenhouse gases. One notable difference between the two control problems is that the technology for the accurate monitoring of many sources and sinks of greenhouse gases has not been developed.

CHOOSING POLICY INSTRUMENTS FOR GLOBAL CLIMATE CHANGE

This chapter focused on evaluating the use of tradeable permits for controlling greenhouse gas emissions. We compared tradeable permits with taxes and with command-and-control regulations, and on the basis of those comparisons, we can offer the following observations: First, international application of effective tradeable permits regimes is likely to be significantly more complicated than domestic applications. Second, given the current understanding of the causes and consequences of global climate change, and limitations on monitoring and enforcement, a comprehensive market-based approach for greenhouse gas emissions is probably not appropriate, even in the United States. Third, domestic application of carbon taxes or tradeable carbon rights for controlling carbon dioxide emissions in the United States is likely to be substantially more cost-effective than command-and-control approaches that achieve similar environmental results.

The best approach to limiting greenhouse gases will probably involve a mixture of instruments.[96] This mix will vary across countries depending on their ability to define baselines and enforce different control strategies. Whereas the United States, for example, might be capable of defining and enforcing a strategy that gives credit for carbon dioxide sequestration from forests, some developing countries might not have the administrative or technical resources.

Theoretically, an international system of tradeable greenhouse gas permits can simultaneously address the issues of cost-effectiveness and equity. As a global problem involving uniformly mixed pollutants, global warming is particularly well suited to an approach that controls carbon dioxide emissions (and perhaps other greenhouse gas emissions) at the aggregate level, while encouraging individual nations with the lowest costs of control to take on added responsibility. Through the initial allocation of permits, questions of fairness between industrialized and developing nations can be addressed directly, and technology transfer can thus be engendered. A key concern in implementing a greenhouse policy—whether it involves tradeable permits or not—will be to ensure that compliance is adequately monitored and enforced.

It is unlikely that scientists, economists, or others will be able to resolve any time soon the myriad uncertainties involved in our current understanding of the causes and consequences of global climate change, but the call for action from some quarters is unmistakable. If governments decide that action is indeed warranted, important policy questions will have to be addressed, possibly quite quickly.

Given the pervasive role of energy generated from fossil fuels in virtually all nations, policies designed to address global climate change could have profound effects on many aspects of our lives. Therefore, at both international and domestic levels, cost-effectiveness must be a central consideration in policy design to ensure that the economic well-being of millions of people around the world is not unduly compromised. A tradeable permit system offers one possible route to a cost-effective attainment of greenhouse gas reduction goals. The potential implementation problems associated with such systems, however, are by no means trivial. Those problems, combined with the lessons learned from the United States's experiences with tradeable permit systems, suggest that careful attention must be given to the design of any system for it to have a reasonable chance of success.

ACKNOWLEDGMENTS

Helpful suggestions on framing the general issues addressed in this chapter were offered by Daniel Dudek, Henry Lee, Bruce Stram, Miriam Avins, and Teresa Johnson. However, we alone are responsible for any remaining errors.

NOTES

1. On 9 May 1992, 143 nations voted at the Earth Summit meeting in Rio de Janeiro to adopt a treaty requiring signatories to limit emissions of greenhouse gases. See William K. Stevens, "143 Lands Adopt Treaty to Cut Emission of Gases," *New York Times*, 10 May 1992, 4.

2. In February of 1992, the United Nations Conference on Trade and Development released a study that proposed a system of tradeable permits to help mitigate climate change, similar to that examined in this chapter. See United Nations Conference on Trade and Development report, "Trading Entitlements to Control Carbon Emissions: A Practical Way to Combat Global Warming" (Geneva: UNCTAD, February 1992).

3. We do not attempt, however, to define a particular allocation of responsibility as equitable or fair. On this issue, see Richard S. Eckaus, "Laissez Faire, Collective Control, or Nationalization of the Global Commons," discussion paper, Center for Energy and Environmental Policy Research, Massachusetts Institute of Technology, 6 October 1992; and H. Peyton Young and Amanda Wolf, "Global Warming Negotiations: Does Fairness Matter?" *The Brookings Review* 10, no. 2 (spring 1992): 46–51.

4. Peter Bohm and Clifford S. Russell, "Comparative Analysis of Alternative Policy Instruments," in *Handbook of Natural Resource and Energy Economics*, vol. 1, eds. Allen V. Kneese and James L. Sweeney (Amsterdam: North-Holland, 1985), 395–460.

5. For a general discussion of instruments for environmental protection, see Robert W. Hahn and Robert N. Stavins, "Market-based Environmental Regulation: A New Era from an Old Idea?" *Ecology Law Quarterly* 18, no. 1 (1991): 1–42.

6. A good example is the control of sulfur dioxide emissions that contribute to acid rain in the United States. It is unlikely that the targeted 10 million tons of reduction would have been agreed to without the introduction of a market-based approach that could achieve the goal using fewer resources than would a command-and-control approach.

7. There are several instruments of importance to policymakers that do not fall conveniently within these two categories (incentive-based and command-and-control), including monitoring and enforcement techniques, use of the courts, and use of information. In designing a system, including one based on economic incentives, these mechanisms should not be overlooked as either complements or substitutes in system design.

8. Regulations usually do not explicitly specify the technology but rather establish standards on the basis of a particular technology.

9. In a survey of eight empirical studies of air pollution control, Tietenberg found that the ratio of the actual, aggregate costs of conventional command-and-control approaches to the aggregate costs of least-cost benchmarks ranged from 1.07 for sulfate emissions in the Los An-

geles area to 22.0 for hydrocarbon emissions at all domestic DuPont plants. See T. H. Tietenberg, *Emissions Trading: An Exercise in Reforming Pollution Policy* (Washington, DC: Resources for the Future, 1985).

10. Numerical examples of the variance of incremental costs of air pollution control are provided by Robert W. Crandall, "The Political Economy of Clean Air: Practical Constraints on White House Review," in *Environmental Policy under Reagan's Executive Order: The Role of Benefit-Cost Analysis*, ed. V. Kerry Smith (Chapel Hill: University of North Carolina Press, 1984), 205–225.

11. For recent theoretical examinations of the effects of alternative environmental policy instruments on the diffusion of new technologies, see Scott R. Milliman and Raymond Price, "Firm Incentives to Promote Technological Change in Pollution Control," *Journal of Environmental Economics and Management* 17 (1989): 247–265; and David A. Malueg, "Emission Credit Trading and the Incentive to Adopt New Pollution Abatement Technology," *Journal of Environmental Economics and Management* 5 (1989): 52–57. For an empirical investigation of these same effects, see Adam B. Jaffe and Robert N. Stavins, "Dynamic Incentives of Environmental Regulation: The Effects of Alternative Policy Instruments on Technology Diffusion," *Journal of Environmental Economics and Management*, forthcoming, Volume 29, No. 1, July 1995. For a discussion of the politics of standard setting and the adverse impacts that command-and-control regulations can have on innovation, see Matt Ridley, "How to Smother Innovation," *Wall Street Journal*, 9 June 1993, A12.

12. Each source's marginal costs of pollution control are the additional or incremental costs for that source to achieve an additional unit of pollution reduction. If these marginal costs of control are not equal across sources, then the same aggregate level of pollution control could be achieved at lower overall cost simply by reallocating the pollution control burden among sources, so that low-cost controllers controlled proportionately more and high-cost controllers controlled proportionately less. Additional savings could theoretically be achieved through such reallocations until marginal costs were identical for all sources.

13. Robert N. Stavins and Bradley W. Whitehead, "The Greening of America's Taxes: Pollution Charges and Environmental Protection" (Washington, DC: Report of Progressive Policy Institute, February 1992).

14. The monitoring burden for both charges and tradeable permits may exceed that required under command-and-control regulations for

political and legal reasons, as well as for technological ones. Environmentalists may require better monitoring approaches in return for allowing environmental agencies to experiment with incentive-based approaches. Moreover, because firms are being charged for their estimated pollution, laws and regulations may require higher levels of monitoring to ensure that charges are neither too high nor too low.

15. The political difficulty of establishing large charges can also be viewed as an advantage of charge systems. Charges tend to give highly visible signals to consumers (and thus to politicians) of the costs of environmental protection and thus encourage the public to consider not only the benefits, but also the costs of environmental protection. For an examination of the magnitude of carbon taxes that would be required to achieve various greenhouse goals, see the chapter by Jorgenson and Wilcoxen in this volume.

16. Robert Hahn and Roger Noll, "Designing a Market for Tradeable Permits," in *Reform of Environmental Regulation*, ed. Wesley Magat (Cambridge, MA: Ballinger, 1982), 119–146.

17. These systems are likely to be the simplest in terms of their monitoring burdens, but can lead to complexities of other kinds. For example, if the environmental concern is carbon dioxide emissions, but the point of control is the carbon content of fossil fuels, it can be desirable to exempt nonfuel uses of petroleum, natural gas, and coal from the system.

18. Exposure trading is discussed by James A. Roumasset and Kirk R. Smith, "Exposure Trading: An Approach to More Efficient Air Pollution Control," *Journal of Environmental Economics and Management* 18 (1990): 276–291. For an examination of risk trading, see Paul R. Portney, "Reforming Environmental Regulation: Three Modest Proposals," *Issues in Science and Technology* 4 (1988): 74–81.

19. In the case of localized air pollution, differences in source location and seasonal factors mean that not all emissions reductions are of equal value in terms of improving air quality, a problem that also applies to command-and-control approaches. But in the case of transboundary pollutants, such as sulfur dioxide as an acid rain precursor, this concern about hot spots is much less important. In the case of global commons problems, such as climate change due to greenhouse gas emissions, this concern is completely irrelevant.

20. For tradeable permit reductions for CFCs, see Robert W. Hahn and Albert M. McGartland, "The Political Economy of Instrument Choice: An Examination of the U.S. Role in Implementing the Montreal Protocol," *Northwestern University Law Review* 83 (1989):

592–611. For point-nonpoint source trading for water quality control, see Henry M. Peskin, "Nonpoint Pollution and National Responsibility," *Resources*, no. 83 (spring 1986): 10–11.

21. An evaluation of the EPA's Emissions Trading Program can be found in Tom Tietenberg, *Emissions Trading: An Exercise in Reforming Pollution Policy* (Washington, DC: Resources for the Future, 1985). For a broader assessment of the EPA's experiences with tradeable permit policies, see Robert W. Hahn, "Economic Prescriptions for Environmental Problems: How the Patient Followed the Doctor's Orders," *Journal of Economic Perspectives* 3 (1989): 95–114.

22. U.S. Environmental Protection Agency, *Emissions Trading Policy Statement*, 51 Fed. Reg. 43,814 (1986) (final policy statement).

23. Richard A. Liroff, *Reforming Air Pollution Regulations: The Toil and Trouble of EPA's Bubble* (Washington, DC: Conservation Foundation, 1986).

24. Daniel J. Dudek and John Palmisano, "Emissions Trading: Why Is This Thoroughbred Hobbled?" *Columbia Journal of Environmental Law* 13 (1988): 217–256.

25. Jeremy Main, "Here Comes the Big New Cleanup," *Fortune*, 21 November 1988, 102–118; and Carolyn Lochhead, "Credit Bartering in the Market for Air Pollution," *Insight*, 3 July 1989, 15–17.

26. Robert W. Hahn and Gordon L. Hester, "Where Did All the Markets Go? An Analysis of EPA's Emissions Trading Program," *Yale Journal of Regulation* 6 (1989): 109–153.

27. U.S. Environmental Protection Agency, *Regulation of Fuel and Fuel Additives*, at 38,078–90 (proposed rule), 49,322–24 (final rule).

28. In each year of the program, more than 60 percent of the lead added to gasoline was associated with traded lead credits. See Robert W. Hahn and Gordon L. Hester, "Marketable Permits: Lessons for Theory and Practice," *Ecology Law Quarterly* 16 (1989): 361–406.

29. The program did experience some relatively minor implementation difficulties related to the importation of leaded fuel. It is not clear that a comparable command-and-control approach would have done better in terms of environmental quality. See U.S. General Accounting Office, *Vehicle Emissions: EPA Program to Assist Leaded-Gasoline Producers Needs Prompt Improvement*, GAO/RCED-86-182 (Washington, DC: U.S. GAO, August 1986).

30. Hahn and Hester, "Marketable Permits."

31. U.S. Environmental Protection Agency, Office of Policy Analysis, *Costs and Benefits of Reducing Lead in Gasoline, Final Regulatory Impact Analysis*. Washington, DC: February 1985.

32. The Montreal Protocol calls for a 50-percent reduction in the production of CFCs from 1986 levels by 1998. In addition, the protocol freezes halon production and consumption at 1986 levels beginning in 1992.

33. The system was designed with allowances to limit both domestic production and consumption. See Hahn and McGartland, "Political Economy of Instrumental Choice."

34. Letter from Richard D. Feldman, U.S. Environmental Protection Agency, 7 January 1991. In addition, there have been a very small number of international trades. Such trading is limited by the Montreal Protocol.

35. As of 1992, no firms were producing CFCs up to their maximum allowable level and permits could not be banked (carried forward). As a result, there is an excess supply of permits. It is possible, however, that there would be an excess supply even if there were no tax and an effective price of zero for permits. This excess in supply is due to firms reacting to changes in regulations and new policy initiatives that call for a more rapid phaseout of CFCs and halons.

36. Clean Air Act Amendments of 1990, Public Law No. 101-549, 104 Statute 2399, 1990.

37. For a description of the legislation, see Brian L. Ferrall, "The Clean Air Act Amendments of 1990 and the Use of Market Forces to Control Sulfur Dioxide Emissions," *Harvard Journal on Legislation* 28 (1991): 235–252.

38. Under specified conditions, utilities that have installed coal scrubbers to reduce emissions can receive two-year extensions of the Phase I deadline plus additional allowances.

39. Utilities that install scrubbers receive bonus allowances if they clean up early. In addition, specified utilities in Ohio, Indiana, and Illinois will receive extra allowances during both phases of the program. All of these extra allowances are essentially tradeable (and bankable) compensation intended to benefit midwestern plants, which presently rely on high-sulfur coal.

40. In general, units with a capacity of 75 MW or more and that emit sulfur dioxide at a rate of 1.2 pounds per million Btu face limitations on total emissions that are related to their actual energy generation during the period 1985–1987.

41. Clean Air Act Amendments of 1990.

42. Daniel J. Dudek, statement made at *Acid Rain: Hearings on S. 1630 Before the Subcommittee on Environmental Protection of the Senate Committee on the Environment and Public Works*, 101st Congress, 1st Session, 1989.

43. For activity of trading among utilities, see, for example, Matthew L. Wald, "Utility Is Selling Right to Pollute," *New York Times*, 12 May 1992, D7; and Matthew L. Wald, "Electric Utility in Ohio to Buy Pollution Rights from Alcoa," *New York Times*, 1 August 1992, D1. For increase in the reduction target, see Dennis Wamstead, "EPRI: SO_2 Model Predicts Trading Market," *Environment Week*, 6 June 1991, 5. For trading on a futures market, see "Chicago Board of Trade to Create Smog Futures," *Washington Post* 17 July 1991, F1; and "Smog Futures to Be Traded," *New York Times*, 22 April 1992, D1.

44. See, for example, Paul L. Joskow, "Implementing the Tradeable Allowance System for Acid-Rain Control" (paper presented at the John F. Kennedy School of Government, Harvard University, 2 October 1991); Robert W. Hahn and Roger G. Noll, "Barriers to Implementing Tradeable Air Pollution Permits: Problems of Regulatory Interactions," *Yale Journal on Regulation* 1 (1983): 63–92; Douglas R. Bohi and Dallas Burtraw, "Utility Investment Behavior and the Emission Trading Market," discussion paper ENR91-04, Resources for the Future, Washington, DC, January 1991; and Robert W. Hahn, "Government Markets and the Theory of the Nth Best," discussion paper 91–14, Center for Science and International Affairs, John F. Kennedy School of Government, Harvard University, December 1991.

45. See Robert W. Hahn and Roger G. Noll, "Environmental Markets in the Year 2000," *Journal of Risk and Uncertainty* 3 (1990): 351–367. For a discussion of some of the trade-offs that may be encountered in the implementation of market-based approaches for limiting greenhouse gas emissions, see Michael A. Toman and Stephen M. Gardiner, "The Limits of Economic Instruments for International Greenhouse Gas Control," discussion paper, Resources for the Future, Washington, DC, December 1991.

46. See, for example, Anne E. Smith, "Issues in Implementing Tradeable Allowances for Greenhouse Gas Emissions" (paper presented at the 84th annual meeting of the Air and Waste Management Association, Vancouver, British Columbia, 16–21 June 1991).

47. For practical policy purposes, acid rain precursors may be thought of as uniformly mixed within relevant regions (airsheds).

48. A comparison of an international system of greenhouse tradeable permits with alternative policy mechanisms is provided by Joshua M. Epstein and Raj Gupta, *Controlling the Greenhouse Effect: Five Global Regimes Compared* (Washington, DC: Brookings Institution, 1990).

49. Alan S. Manne and Richard G. Richels, "International Trade in

Carbon Emission Rights: A Decomposition Procedure," *American Economic Review Papers and Proceedings* 81 (1991): 135–139.

50. For example, suppose Holland imposes a carbon tax that is higher than the equilibrium price of carbon permits in an international trading regime. Then, marginal control costs for limiting carbon would vary across nations, and some of the expected cost savings from an economic incentive approach would not be realized. Vice President Gore has, in fact, suggested a similar approach for the United States. See Albert Gore, *Earth in the Balance: Ecology and the Human Spirit* (Boston: Houghton Mifflin, 1992).

51. This set of conditions for a successful tradeable permit market was originally developed in the context of newsprint recycling credits by Terry M. Dinan, "Implementation Issues for Marketable Permits: A Case Study of Newsprint," *Journal of Regulatory Economics* 4 (1992): 71–87.

52. Furthermore, in the presence of transaction costs, the initial allocation of permits can affect the amount of trading that will occur, the equilibrium allocation, and the aggregate costs of control. See Robert N. Stavins, "Transaction Costs and Tradeable Permits." *Journal of Environmental Economics and Management,* forthcoming, Vol. 29, No. 1, July 1995.

53. Brokers have been successful in lowering transaction costs in other tradeable permit markets, such as for lead and criteria air pollutants. See Dudek and Palmisano, "Emissions Trading."

54. On the other hand, carbon dioxide emissions are correlated with the emissions of other pollutants that can have localized effects, including sulfur dioxide.

55. Stavins in "Transaction Costs" provides an examination of the implications for public policy of transactions costs in tradeable permit markets in terms of choosing between command-and-control and market-based instruments, choosing between tradeable permits and pollution taxes, and designing tradeable permit systems.

56. Robert W. Hahn, "Market Power and Transferable Property Rights," *Quarterly Journal of Economics* 99 (Nov. 1984): 753–765.

57. Such manipulation is unlikely if there are a large number of buyers and sellers in the market.

58. See, for example, Daniel Dudek and Alice LeBlanc, "Offsetting New CO_2 Emissions: A First Rational Greenhouse Policy Step," *Contemporary Policy Issues* 8, no. 3 (1990): 29–42.

59. See, for example, Jonathan Green and Philippe Sands, "Establishing an International System for Trading Pollution Rights," *International Environmental Reporter,* 12 February 1992, 80–85.

60. Within the general notion of a carbon-rights system among producers, there also exist a number of potential levels of trading: (1) primary energy extractors and importers, (2) energy processors, and (3) energy distributors. See Anne E. Smith, Anders R. Gjerde, Lynn I. DeLain, and Ray R. Zhang, "CO_2 Trading Issues; vol. 2., Choosing the Market Level for Trading" (Washington, DC: Decision Focus Incorporated, May 1992).

61. Daniel J. Dudek and Alice LeBlanc, "Preserving Tropical Forests and Climate: The Role of Trees in Greenhouse Gas Emissions Trading" (New York: Environmental Defense Fund, February 1992).

62. Quantitatively defining the magnitude, size, and rates of the loss of sinks, such as forests, would be an enormous undertaking. Satellite monitoring is a critical tool. This effort must be made, however, whether or not the international trading of credits involving sinks is authorized, so long as carbon dioxide emissions from destruction or creation of such sinks is incorporated into a convention.

63. See, for example, Marlise Simons, "North-South Divide Is Marring Environment Talks," *New York Times*, 17 March 1992, A8.

64. R. A. Houghton, "A Blueprint for Monitoring the Emissions of Carbon Dioxide and Other Greenhouse Gases from Tropical Deforestation" (paper presented at the Woods Hole Research Center, Woods Hole, MA, 20 January 1992).

65. For further discussion of the arguments in favor of the multiple-gas approach, see Richard B. Stewart and Jonathan B. Wiener, "The Comprehensive Approach to Global Climate Policy: Issues of Design and Practicality," *Arizona Journal of International and Comparative Law* 9 (1992): 85–113.

66. Arguments against the multiple-gas approach are found in David G. Victor, "Limits of Market-based Strategies for Slowing Global Warming: The Case of Tradeable Permits," *Policy Sciences* 24 (1991): 199–222.

67. Alexander Cristofaro and Joel D. Scheraga, "Policy Implications of a Comprehensive Greenhouse Gas Budget," working paper, Office of Policy Analysis, Office of Policy, Planning, and Evaluation, U.S. EPA, Washington, DC, September 1990.

68. Easing more gases into the trading framework may be easier said than done, particularly if trading is allowed across pollutants. For example, methane being introduced after a decade of carbon dioxide trading could have major effects on participants in the carbon dioxide market, giving rise to political resistance if the addition of methane resulted in an effective devaluation of carbon dioxide permits.

69. A recent example of such cooperation is the toxicological testing of alternatives to CFCs that was funded by several major chemical firms.

70. While the initial goal can be stated in many forms, it is critical that it be sufficiently flexible to accommodate trading. For example, restricting the ambient concentrations of each greenhouse gas to a fixed level would place ultimate limits on trading among gases. On the other hand, a goal stated in terms of maximum allowable temperature change could be translated into a carbon dioxide equivalent if carbon dioxide is the numeraire gas.

71. Michael Grubb and James K. Sebenius, "Participation, Allocation, and Adaptability in International Tradeable Emission Permit Systems for Greenhouse Gas Control," in *Proceedings of OECD Workshop on Tradeable Emission Permits to Reduce Greenhouse Gases* (Paris: OECD, June 1991), and Geoffrey Bertram, "Tradeable Emission Permits and the Control of Greenhouse Gases," *The Journal of Development Studies* 28 (1992): 423–446.

72. For example, under an allocation system related to population levels, the big players in the market would likely be India and China, as permit sellers, and the United States and perhaps the former Soviet Union, as buyers. See Epstein and Gupta, *Controlling the Greenhouse Effect*.

73. For example, the Canadians proposed the use of population and GNP combined as allocation criteria when CFC reduction obligations were being considered in the development of the Montreal Protocol.

74. This statement is an abstraction from the potential consequences of significant transactions costs. See discussion above.

75. Most proposals for allocating control obligations among nations call for proportionately higher rates of reduction in emissions by industrialized countries (and, among the industrialized countries, by the United States) and for substantial reductions in the predicted rates of increase in carbon dioxide emissions by most developing countries. See, for example, Florentin Krause, *Energy Policy in the Greenhouse: From Warming Fate to Warming Limit—Benchmarks for a Global Climate Convention* (The Hague: Dutch Ministry of Housing, Physical Planning, and Environment and the European Environmental Bureau, 1989); Christopher Flavin, "Slowing Global Warming: A Worldwide Strategy," Worldwatch Paper 91 (Washington, DC: Worldwatch Institute, October 1989); and David Wirth and Daniel Lashof, "Beyond Vienna and Montreal—Multilateral Agreements on Greenhouse Gases," *Ambio* 19 (1990): 305–310.

76. The chapter by Norberg-Bohm and Hart in this volume discusses

the barriers to successful technology adoption in developing countries.

77. For further discussion, see James T. B. Tripp and Daniel J. Dudek, "Comments on the IPCC Working Group III Economic Measures Paper" (New York: Environmental Defense Fund, January 1990). For examinations of the international distributional implications of carbon dioxide controls, see Alan S. Manne and Richard G. Richels, "Global CO_2 Emission Reductions: The Impacts of Rising Energy Costs," *The Energy Journal* 12, no. 1 (1991): 87–107; and David Pearce and Edward Barbier, "The Greenhouse Effect: A View from Europe," *The Energy Journal* 12, no. 1 (1991): 147–160.

78. Scott Barrett, "Economic Instruments for Global Climate Change Policy," working paper, London Business School, November 1991.

79. Jose Goldemberg, "Brazil's Small Share of the Greenhouse," *New York Times*, 28 July 1990, A20.

80. This group would likely have limited authority, given the interest of countries in maintaining their sovereignty. The group could be patterned after the Montreal Protocol, which has designated an implementation committee consisting of six parties. This committee assesses the extent of compliance and works with countries to help bring them into compliance.

81. A variant on the preceding proposal would be to allow selected countries to trade internationally and allow the supranational authority to review special requests from other countries on a case-by-case basis. This alternative would be administratively cumbersome.

82. David Pearce, "Greenhouse Gas Agreements: Internationally Tradeable Greenhouse Gas Permits," working paper, Department of Economics, University College, London, March 1990.

83. Robert W. Hahn and Gordon L. Hester, "The Market for Bads: EPA's Experience with Emissions Trading," *Regulation* 3/4 (1987): 48–53.

84. See David Victor, "Limits of Market-based Strategies." The problem of defining which gases are appropriate for economic incentives is by no means straightforward. Consider the case of CFCs as they relate to climate change. CFCs were thought to have the greatest influence on climate change on a mass basis; however, CFCs are now considered to have a much lower impact on climate, and may even have a negligible one. Of course, CFCs may need to be limited because of their effect on stratospheric ozone, but this example illustrates the potential difficulties of defining control strategies when the science is uncertain.

85. See, for example, the chapter by Stram in this volume; Dale W. Jorgenson, Daniel T. Slesnick, and Peter J. Wilcoxen, "Carbon Taxes

and Economic Welfare," in *Brookings Papers: Microeconomics 1992* (Washington, DC: Brookings Institution), 393–454; U.S. Congressional Budget Office, *Carbon Charges as a Response to Global Warming: The Effects of Taxing Fossil Fuels*, Washington, DC, August 1990; David Pearce, "The Role of Carbon Taxes in Adjusting to Global Warming," *The Economic Journal* 101 (1991): 938–948; and Lawrence H. Goulder, "Effects of Carbon Taxes in an Economy with Prior Tax Distortions: An Intertemporal General Equilibrium Analysis," working paper, Stanford University, January 1993.

86. Compared with conventional command-and-control regulations, both charges and permits provide an explicit price signal about the marginal cost of limiting emissions.

87. The U.S. federal budget negotiations of 1990 provide at least indirect evidence of this point. The Bush administration proposed a 25 cent per gallon gasoline tax increase; Congress eventually enacted a 5 cent increase. The demise in 1993 of the Clinton administration's proposed Btu tax is a more recent example.

88. These distinctions between resource transfers assume that permits are distributed free of charge to firms, not auctioned by the government. In the latter case, permits and charges are quite similar in terms of these financial transfers. Moreover, it is possible to design a subsidy or charge scheme that is similar in distributional terms to a permit approach.

89. Reference here is to environmental problems with highly nonlinear dose-response functions, that is, threshold impacts.

90. Martin L. Weitzman, "Prices vs. Quantities," *Review of Economic Studies* 41 (1974): 477–491.

91. This change can be offset to some degree by expanded production that results from lower total operating costs.

92. Although there is no cutoff point, it is unlikely that firms or nations could engage in price-setting behavior if they controlled less than 10 percent of the market. See F. M. Scherer, *Industrial Market Structure and Economic Performance* (Chicago: Rand McNally, 1980). Ultimately, the question is whether other firms present credible threats of entry to the market, that is, whether the market is contestable. If so, it is less likely that anticompetitive behavior can thrive. See William J. Baumol, John Panzar, and Robert Willig, *Contestable Markets and the Theory of Industrial Structure* (New York: Harcourt Brace Jovanovich, 1982).

93. Stavins, "Transaction Costs."

94. Geoffrey Heal, "International Negotiations on Emission Control"

(paper presented at the National Bureau of Economic Research, Cambridge, MA, August 1991).

95. The likelihood of reaching an international environmental agreement will be affected by several factors. See James K. Sebenius, "Designing Negotiations toward a New Regime: The Case of Global Warming," *International Security* 15 (1991): 110–148; and Robert W. Hahn and Kenneth R. Richards, "The Internationalization of Environmental Regulation," *Harvard International Law Journal* 30 (1989): 421–446.

96. Rosina Bierbaum and Robert M. Friedman, "The Road to Reduced Carbon Emissions," *Issues in Science and Technology* 8 (1992): 58–65.

7 A Carbon Tax Strategy for Global Climate Change

■ **BRUCE N. STRAM**
Enron Corporation
Houston, Texas

Any policy that can prevent global warming will substantially burden all nations' economies, yet it must win their voluntary support. This situation requires that the policy minimize the costs of ameliorating global warming and, in particular, provide incentives to developing countries to lower their greenhouse gas emissions. In addition, the policy must be flexible enough to respond to the growing level of scientific knowledge concerning global warming. To craft a policy that satisfies all these requirements is indeed a daunting task.

Unfortunately, this task cannot be approached with a quiver full of successful policies tested in past endeavors. In recent decades, particularly in the United States, environmental and energy policies frequently have been costly, ineffective, or both. One common approach has been to prescribe abatement technologies or procedures that firms must follow. However, these policies have tended to fail because firms do not have the flexibility to choose the best pollution reduction alternatives and have no incentives to search for more effective means of remediation. Recently, environmental regulators at all levels of government have expressed strong interest in the use of market-based mechanisms that dramatically change the incentives to reduce pollution.

This chapter explores how a system of uniform carbon taxes in developed nations, coupled with subsidies for the abatement of emissions in developing countries, could facilitate worldwide reduction in greenhouse gas emissions. Such a mechanism could provide the impetus to maintain international cooperation: developed nations would charge uniform taxes and would award subsidies to reduce the effects of global warming, whereas developing nations should respond to the direct subsidy incentive. It should also be technically easy and politically less difficult to expand such a policy to include other greenhouse gases and to adjust the taxation and subsidy levels as new information accumulates. Perhaps the most beneficial aspect of this proposed plan is that it would encourage technology innovations rather than provide a perverse incentive for the stagnation of technology development.

In this chapter, I review the Clean Air Act to show how the traditional approach of mandating specific abatement technologies is counterproductive, and I discuss recent experiments with tradeable permits—a market-based mechanism that has many of the same virtues as a pollution tax. A key portion of the chapter illustrates how a carbon tax can be adapted to be the core of a truly global strategy that has administrative and policy advantages over other approaches.

THE UNITED STATES'S EXPERIENCE: THE CLEAN AIR ACT AS A CASE STUDY

Controlling greenhouse gas emissions will require a substantial intervention into the world economy. This intervention must successfully introduce pollution abatement technologies, dramatically change how the United States and other countries consume energy, or both. However, the track record of the U.S. government's policies involving such intervention has been particularly weak. Recent attempts at regulation have required similar, but smaller, interventions than will likely be required to reduce carbon dioxide emissions. The history of the Clean Air Act provides an illustration of how command-and-control intervention can actually be counterproductive. In 1990 Congress, realizing the relative ineffectiveness of command-and-control regulation, amended the act and introduced an array of incentive mechanisms that promise to accelerate the nation's efforts to reduce carbon dioxide and nitrogen oxides emissions.

In the late 1960s, an explosion of attention to the environment led Congress to enact the Clean Air Act of 1970. While the act's goals were laudatory, the means mandated to achieve those goals have been comparatively ineffective. The act sought to protect the health of even the most sensitive people from seven airborne pollutants without regard to cost. The principal pollutants controlled by the act are sulfur oxides, particulates, ozone, carbon monoxide, and nitrogen oxides.

The Clean Air Act is a framework; which pollutants are to be controlled and the goals to be achieved are identified, but many of the detailed decisions about how to reduce pollution are left to the U.S. EPA and state authorities. The EPA determined ambient air quality standards for the pollutants listed in the bill, and state agencies were charged with establishing plans for controlling emissions. In many regions, the emissions targets were not met.

1977 AMENDMENT TO THE CLEAN AIR ACT

In 1977 amendments to the Clean Air Act extended the deadlines and designated those areas that failed to meet the standards as nonattainment areas. These areas suffered particularly stringent restrictions on new pollution sources that were (and still are) required to maintain the lowest achievable emission rate (LAER), as specified in each state's plan.

The amended act also required that even in areas meeting the standard, new sources must use the best available control technology (BACT) in order to prevent significant deterioration of air quality. Existing sources of pollutants remain subject to the frequently more lenient standards of "reasonably available control technology" or the original state plans. The states were to implement LAER and BACT standards, but the EPA established new source performance guidelines to help the states in their efforts. The states were (and are) ultimately responsible for attaining acceptable levels of air quality; if they are not in attainment, states must convince the EPA that their regulations will generate reasonable progress toward attainment.

Thus, at the end of the 1970s, the air pollution policy had two main features: a policy that grandfathered most preexisting sources (that is, subjected them to the older, less-stringent standards) and a set of government-specified abatement technologies for new sources. The term "command-and-control" has been coined to describe this type of regulatory methodology (Tietenberg 1986: 273–277). The specification of technological standards for new sources meant that there was little economic advantage in using low-sulfur coal; new users of low-sulfur coal were still required to install expensive state-of-the-art abatement equipment. The Clean Air Act rules similarly offered little incentive to switch to cleaner fuels; prescribed abatement technology was defined by types of fuel in use. The older plants, most of which were effectively grandfathered, were expected gradually to be phased out, leaving only new sources, which would all meet performance standards.

This approach reflects a political compromise between competing interests. On one side were the coal-using or producing industries, such as electric power generation companies and their associated labor organizations. On the other side were environmental groups. The specification of particular technologies was not mandated by the Clean Air Act but was driven by political compromise at both congressional and regulatory policy levels.

The compromise made sense for both sides. The notion of new source performance standards gave a great deal to industry and labor; it preserved substantial property values and thousands of jobs. Environmentalists gained in two ways. First, they achieved clear results—changing the technology used by virtually every utility and large user of coal in the country was a major accomplishment. Second, there was also a more subtle, political advantage. Environmentalists' political strength ultimately lies in their representation of the public's strong desire for a clean environment. But while everyone is in favor of a better environment, the issue is more divisive when one asks, "How much are we willing to pay?" Citizens may become less committed when the cost of environmental protection must be borne in their own salaries and jobs and is reflected in their utility bills (Cambridge Energy Research Associates 1992). A focus on new source performance standards neatly dodges the issue: the immediate impact is small, and the long-term costs build slowly and are buried in the cost of new plants. Few jobs are visibly lost immediately, few plants lose value or are shut down, and few consumers immediately pay utility bills that include environmental charges.

Unfortunately, the compromise does not work well as a policy because the economic incentives are wrong. The grandfathered power plants and industrial boilers have a competitive advantage over new plants with expensive abatement equipment. Rather than fade away, the older plants have tended to stay in operation much longer than expected, emitting huge quantities of sulfur oxides, nitrogen oxides, particulates, and other pollutants.

1990 AMENDMENTS TO THE CLEAN AIR ACT

The poor performance of command-and-control emissions regulation led Congress to consider alternative approaches to convince firms to reduce air pollution. One of these is emissions-credits trading. Under this system, plants and factories are permitted to emit pollutants as long as they arrange a reduction in emissions from another source within the defined area. In effect, such an arrangement requires an emissions permit for operation, so that new sources are willing to pay. Environmental authorities may well help arrange the trades. A principal reason why this system makes sense is that many grandfathered polluters, who have had little incentive to reduce emissions, can in fact do so at a relatively low cost. New businesses bribe old businesses to cut pollution, thus providing for the economic incentives lacking in the old provisions of the act (Tietenberg 1986: 280–295; see also the chapter in this volume by Hahn and Stavins).

Sulfur dioxide emissions trading is a centerpiece of the amendment to the Clean Air Act passed in 1990. Under this program, permits are issued to existing polluters that specify how much sulfur dioxide they have a right to emit. New and existing businesses are allowed to purchase these rights from polluters that emit less than their permitted limit. Thus, low-cost abaters sell permits to high-cost abaters, and new businesses that pollute must purchase them to operate.

This framework provides the proper incentive at the margin to induce polluters to minimize the costs of achieving the targeted level of abatement and provides an incentive for technical innovation. However, this approach may be inferior to a tax regime when applied to the global warming problem.

EMISSIONS TAXES

Economic theory predicts that the free interplay of market forces—each participant pursuing his or her own self-interest—leads to an aggregate result that minimizes misdirected activity or waste. However, this theory assumes that all economic actors bear all of the costs associated with pursuing a particular activity. Pollutant emission imposes costs on others who are innocent of the activity; the economic paradigm of efficiency breaks down when such externalities exist.

A tax on pollutant emissions equal to the societal costs of the pollutants reestablishes efficiency; an effective pollution tax creates the incentive to polluters to limit their emissions. For instance, if the tax is a charge per unit of emissions, then by reducing emissions, polluters reduce their tax payments. Polluters are thus willing to make expenditures on abatement technology in order to reduce costs. Ideally, businesses reduce pollutant emissions up to the point at which the additional cost of abating a quantity of pollutant equals the tax on that quantity. If the tax is set at the right level—that is, equal to the costs imposed on the rest of society by pollutant emission—the polluter should make the proper decision as to abatement expenditures. Furthermore, the tax should strongly encourage technological innovation in treatment or production methods that are less polluting. Polluters will then have an incentive to search for the most efficient means to reduce their emissions, thus providing a dynamic market for new abatement technologies and options.

In contrast, a general technology standard is likely to establish levels of abatement that are inappropriate for a wide range of circumstances, likely requiring either too little or too much abatement. The polluting industries have a perverse incentive to discourage innovation in abatement technology, since evolution of better technology might lead to regulatory standardization of more expensive technology.

Emissions taxes share many properties with emissions-credit trading, in which pollution credits are traded in the marketplace and the price of pollution credits creates a strong incentive for every polluter to abate regardless of the initial allocation of rights. Those with excess rights can sell them. Saleable emissions rights also create an incentive for technology development similar to that in a tax regime.

A CARBON TAX AS A FRAMEWORK FOR GLOBAL WARMING POLICY

Theoretically, an equivalent tax could be calculated for any emissions-trading target, and vice versa. However, it is my view that emissions trading is inferior to taxation for facilitating domestic political decision making, international cooperation, and administrative ease. Pollutants entitlement would create a stronger vesting of interests in the level of the abatement targets than would a tax. One can expect these perverse interests to be expressed in the political process. This section outlines a possible approach to the reduction of greenhouse gas emissions on the basis of carbon taxes. First, I sketch a proposal, or working hypothesis; then I compare this hypothesis to systems based on permit and command-and-control regimes.

A carbon tax would require the OECD nations to agree that a tax mechanism should be the principal means of controlling greenhouse effects. Initially, a study would need to be commissioned from a multilateral task force on the possible model tax structure, after which the OECD nations would agree to a uniform carbon dioxide tax level, which would be implemented individually by each nation.

An appropriate portion of the tax proceeds would be transferred by each country to a hypothetical International Environment Fund (IEF), administered either by an existing institution, such as the World Bank or the IMF, or by an agency created for this purpose. The IEF, in consultation with the OECD nations, would design and develop a program of incentives to stimulate developing countries to reduce carbon dioxide emissions. In addition, the IEF would fund global warming research to be conducted under the auspices of an existing international research body.[1]

As our knowledge base increases, the OECD nations would adapt the tax mechanism to include other greenhouse gases. Periodically, if the evidence merited, they would also adjust the tax level.

POLITICAL DECISION MAKING: ENVIRONMENT VERSUS COST

Scientific expertise currently falls far short in its ability to determine how much damage global warming gases are likely to cause and, hence, of being able to specify an appropriate level for an emissions tax. However, a tax is most useful precisely because the experts are uncertain. The critical environmental question is not whether we value the environment, but how much. What sacrifices in comfort, convenience, and even necessities are citizens willing to make to preserve their habitat? The impact of an emissions tax on individuals will be made clear during debate. For citizens (or their representatives) to weigh the uncertain evidence and vote on such a tax is the best measure of how much society values reducing the risk of climate change. The public, as with scientists and environmentalists, will be unsure about the pattern of impacts and the marginal cost of emissions, but in a tax regime people will know how much the policy will cost and can indicate whether they are willing to pay for it. In contrast, past command-and-control regulation efforts have obscured rather than clarified the central issue of cost versus environment. With command-and-control regulations, the potential cost to individuals is as uncertain as the science. Even today, it is not known how much past command-and-control policies have cost. The only clarity achieved with command-and-control systems is that people understand what abatement procedures will be followed, although this clarity is likely to be lost in implementation. They do not, however, understand the impact on emissions, the benefits from reduction, or the cost to society. Weighing specified procedures against concern for the environment is neither revealing nor useful. The clearest focal point—the form of abatement procedures—is virtually meaningless to the electorate.

Emissions trading, on the other hand, establishes with higher certainty exactly what level of emissions will occur, leaving significant uncertainty as to the costs and benefits from the level of abatement that is implicit in the target level of emissions. The quantity of emissions reduction is clearly a useful piece of information on which the decision process can focus; however, environmentalists' desire to limit emissions may outrun the public's willingness to pay, leading, perhaps, to future policy gridlock.

COOPERATION AMONG DEVELOPED NATIONS

An environmental tax can provide a useful focal point for the OECD nations. Relative to command-and-control approaches, an environmental tax is competitively neutral. That is, nations suffer cost and competitive harm from a tax in proportion to the damage their emissions cause. Countries that contribute less to the problem will suffer less economic harm. Those that have already undertaken abatement efforts will also suffer less additional cost. Obviously, this version of equity invokes a normative principle (that is, the polluter pays), which is not universally accepted, particularly by the losers. But the imposition of costs to those whose activities cause harm is an easily understandable concept that has broad appeal; it is, at least, a weapon that can be wielded when negotiations break down because losers are intransigent.

Alternative emissions-control policies, on the other hand, would obscure equity comparisons. Consider the problem of implementing command-and-control policies worldwide. If each country develops its own control regulations, different production methods, and perhaps different abatement technologies, the level of effort in each nation must be compared and then adjusted to achieve a sense of equity. Alternatively, international negotiations could attempt to set up parallel regimes in every nation. Either course provides a wide venue for both honest and self-serving differences in each nation's views of basic facts. Suppose, for example, that the United States required that 50 percent of all new electricity-generating facilities use natural gas instead of coal, whereas Germany requires that 25 percent of new power plants capture carbon dioxide emissions for disposal. How is one to estimate whether these constitute equal efforts toward abatement? By emissions avoided per capita? Per unit of GNP? By expenditure per unit of emissions avoided? It could even be argued that the United States is not reducing emissions on the grounds that, based on economic considerations, 50 percent of the new generating capacity would have been gas-fired even without controls.

Establishing emissions-trading targets presents problems as well. Permitted emissions quantities must be allocated among nations (and ultimately among different polluters) in order to implement an emissions-trading system. Those who get more win; those who get less lose. Setting the allocations at current levels creates a disadvantage for those nations that have already made reduction efforts or that have low GNP per capita. Is the numeraire of target setting to be emissions per capita, per unit of GNP, or per unit of abatement expenditure? The negotiations leading up to the Earth Summit and those taking place within the EC have demonstrated the difficulty of such a target-setting exercise. Setting targets for developing countries could be even more painful.[2]

If environmental taxes are equalized, however, everyone is equal in one sense: the industries of each country will pay a tax equal to the agreed-upon costs of the emissions they cause. Furthermore, each country's industries will have the same incentive to abate their greenhouse gas emissions.

FUNDING CLEAN TECHNOLOGY TRANSFER

The proposal outlined here uses a portion of the revenue from the carbon tax to fund an IEF. The primary purpose of the fund would be to support the efforts of developing and Eastern Bloc nations in reducing greenhouse emissions. The support could be in the form of a direct payment for emissions reductions that was set (as a maximum) equal to the level of reductions multiplied by the OECD tax. That is, if the OECD tax is $5 per ton of carbon dioxide, then developing nations would receive a payment of an equivalent amount for every ton they abated. The support might be paid directly to governments to reward them for reducing emissions; it could also be paid to affected industries on the basis of emissions reduction. The fund would provide developing and Eastern Bloc nations with an incentive to reduce emissions without that effort imposing an undue cost to their economies.

One obstacle to implementation is that a payment incentive mechanism is more complex than a tax. A tax system requires only the imposition of a tax per unit of pollutant emitted, which simply requires the measurement of emissions levels. The means of measurement, both direct and indirect, become ever more feasible. To calculate an equivalent payment, an emissions baseline must be established for each polluter so that the polluter is paid for the amount that emissions are reduced below the target level. In a static world, such a system would be relatively simple to implement; however, in a world of economic growth and technological change, the baseline is continuously changing. Any payment should provide an incentive to reduce emissions below the level that would have otherwise occurred, but defining "what would have otherwise occurred" is difficult.

These complications are similar to the difficulties that would likely arise from setting national emissions targets for emissions-trading regimes. However, there is a critical difference: an IEF would not need consensus among the developing nations to achieve results. It could proceed by way of bilateral agreements. Furthermore, the authority administering the fund would have the upper hand in any negotiation, for it would have the power of the purse. The authority could simply assign a baseline and make an offer. When reductions occur (or are judged to occur), payments are made. If reductions stop, payments stop. Where economies are relatively open, the authority could even approach industry without dealing with governments.

Setting such targets would require a detailed and sophisticated knowledge of economies around the world, as well as of appropriate technologies. Fortunately, such expertise exists in the World Bank, the IMF, and the UN, whose economic development activities have become increasingly sophisticated. Each organization has observed and participated in a variety of development projects, some of which have worked and some of which have not. They have learned through failure and have become much more market-oriented in their approach to projects.

One of the reasons why development projects fail is that the successful transfer of new and sophisticated technologies to Third World environments is very difficult to achieve.[3] Since new, efficient technologies generate fewer greenhouse gases, a Third World emissions reduction effort may involve substantial technology transfer even at its inception. The World Bank, the UN, and the IMF have all succeeded and failed in achieving technology transfer.[4] This experience could be put to good use. Others have observed that international funding for such technology transfer has been relatively paltry to date. A key advantage of an environmental tax is that it would provide an immediate and substantial fund for this purpose. The fund should not, however, be in the business of selecting clean technologies; rather, it should provide an incentive to implement cleaner technology. Technology assessment comes into play only through setting an emissions baseline to credit reductions and to assess (in the absence of verifiable measurement) whether reductions have, in fact, occurred.

The command-and-control approach, on the other hand, offers no support for technology transfer to developing countries. It is difficult to imagine that developing countries will voluntarily choose to burden their economies to better the world environment. Typically, developed nations did not make such a choice, but pursued development and ignored environmental consequences. Support for the compliance of developing countries could be provided from general revenues, but it is difficult to imagine that the developed world, while agreeing to burden its own economies, would simultaneously agree to support developing countries.

An emissions-trading system, however, could be structured to provide such funding. A disproportionate share of emissions permits (of the total global emissions target) could be set aside to support the compliance of developing nations. These permits could be sold to provide a fund parallel to the one proposed above. Alternatively, they could be allocated to developing countries for distribution to industry or for sale in return for meeting compliance targets. However, even an intentionally disproportionate allocation of credits is likely to be viewed as inadequate by developing countries. As with anyone else, developing countries will prove to be very imaginative in determining their just desserts. Furthermore, the award of emissions credits would tend to be an *ex ante* reward to governments, while a subsidy would tend to be a contemporaneous or *ex post* reward for performance.

WEALTH TRANSFER: THE HIDDEN AGENDA

All strategies of pollution abatement assume that there is some level of emissions below which it is too costly to achieve further reductions. An emissions tax equal to incremental global damages caused by those emissions would confront individual economic actors with the key elements of the societal choice—the costs of abatement versus the damages from emissions—and they would treat the matter like any other economic choice. Firms pay to pollute. Similarly, tradeable emissions permits create saleable rights to pollute. Both approaches raise an issue that plagues environmental policy making: if allowing emissions creates an economic value much like the value of a natural resource, who gets the wealth? An emissions-trading regime explicitly creates this wealth and directs it to industry or government. A pure tax regime collects the wealth as tax receipts.

A global emissions-trading system requires that each nation prevent economic actors under its jurisdiction from emitting pollutants without a permit. Permits will quickly, if not immediately, become thought of as a wealth equivalent within any negotiation. The total of participants' expectations of what their shares should be will greatly exceed the value of the pool.

In comparison, the proposed environmental tax creates two pools of value: tax receipts and emissions-reduction subsidies. Each OECD nation receives the tax receipts from its own carbon tax. No further negotiations are required to divide those receipts among them; there is no expectation or precedent that one nation should have a claim on the tax receipts of another. Generally speaking, environmental tax receipts will be greater per capita in nations with greater emissions-reduction burdens and thus will be available in rough proportion to the need to ameliorate economic harm. This distribution is a reasonably equitable one.

The IEF's mission would be to transfer some part of this wealth to developing countries by establishing emissions targets and by paying for reductions below the target. Obviously, developing countries would prefer to negotiate larger rather than smaller targets. In the end, however, since the target represents a subsidy, any nation that fails to agree simply harms itself.

If developing countries enjoy greater economic growth relative to developed nations, their emissions targets might need to be raised (depending on the evolution of technology), thereby placing a greater burden on IEF funds. But this burden would be limited: a growing developing country eventually becomes a developed nation, which ends eligibility for the subsidy. The key issue to be addressed here is how a struggle over wealth distribution may hinder the development of a truly global climate change policy and how policies might mitigate such effects.

ADMINISTRATIVE FLEXIBILITY AND ADAPTABILITY

The accumulation of scientific knowledge is likely to change views of the dangers of potential warming effects and of the relative impacts of different gases. An environmental tax regime could be adapted to such change by a single procedural step within an existing administrative structure. Countries need only agree on adjustments to the tax—either adjustments to the overall level, if the accepted view of global warming damages has changed, or for a particular pollutant, if the estimated impact has changed. Adapting to the changed tax level would be left to individual businesses within each economy. Abatement subsidies for developing countries would also need to be adjusted.

However, an emissions-trading system would require resetting the global target, either for overall emissions or for a particular pollutant, as well as reallocating targets among countries and industries. This extra procedural step might be difficult. If, for example, targets were tightened, which is parallel to raising taxes, there would be winners and losers. Consequently, agreement to reallocate permit quotas would be very difficult and could lead to a serious breakdown in negotiations. Even if permit allocations were reduced, some companies, and even countries, could be absolute winners if the increase in the price of emissions permits more than offset the reduction in the quantity available to them. The potential existence of absolute wins, particularly since they are windfalls, will result in continued criticism that the system is inequitable. In any case, there is no reason to suppose that additional rounds of wealth redistribution, in the form of allocating additional permits or reducing those in circulation, will be easier than the first.

The adaptation of command-and-control strategies to changes in scientific understanding would be even more difficult. As illustrated by legislative debates on the amendments to the Clean Air Act, readjusting environmental legislation requires laborious renegotiating over what constitutes appropriate abatement technologies. If this were ever accomplished once on a worldwide scale, who would want to do it the second, third, or fourth time?

POLITICAL FEASIBILITY

Many of the advantages of an emissions tax system have been known for some time, yet such a system has not been used. Why? Could a tax approach be adopted now? An emissions tax policy is thought to be beneficial because it internalizes external costs and microeconomic decision making is decentralized, the same advantages that explain the superior performance of market-oriented economies compared with centrally directed ones. In the United States, however, market-based approaches depend on the decisions of a disparate group of decisionmakers—thousands of energy-using consumers—and thus political players view the results as uncertain.

An economist would contend that results are certain: if the tax level is correct, then the level of abatement will be correct, and thus the impact on the rest of the economy will be the least costly possible and the net benefit will be as great as possible. However, interest groups involved in the political debate view the outcome as uncertain, and for each of them it is. The electric power company asks what will happen to its plants and its shareholders. The economist doesn't know and doesn't think it matters to the plan as a whole. The environmentalist asks how the pollutants will be abated and by how much emissions will be reduced. The economist replies that the pollutant may be abated in some instances and not in others, and by exactly which means and by how much is uncertain. The economist is certain only that the outcome will yield the best results at the lowest cost.

Such explanations provide little comfort to interest groups. If a tax is seriously considered, each group will estimate the consequences of the tax and will assume the worst outcome for itself—that is, industry will assume that costs and competitive impact will be worse than is likely, and environmentalists will assume that emissions will be greater than is likely. These natural assumptions make compromise harder.

Command-and-control policy options look more attractive because they appear to provide relatively detailed plans (for example, what new equipment will be installed and by whom). Thus, political compromise is often fostered. Of course, these specific plans will inevitably have significant unintended consequences and are not nearly so certain as their precision leads proponents to believe.

Given this situation, how can one expect that a tax, whatever its advantages, could pass? One could appeal to the players to submerge their individual fears and concerns in order to promote the common good. More important, the potential scale and scope of the global warming problem should change the players' incentives. Global warming has a potential impact one or two orders of magnitude larger than any environmental issue considered previously. Many more players, both on the industrial side and the environmental side, will be involved. From industry's point of view, the command-and-control approach tends to lock it into a regulatory process; however, a tax provides a framework for firms to address pollution as a business problem. In the past an industry group has "won," within the command-and-control framework, by gaining a legislative preference under the regulations. Within an incentive system, the industry (or individual business) gains by making good choices about managing its emissions. A tax or credit system makes the choice between rewarding good lobbying or good business decisions much more transparent. The problem is large enough that industry has a collective interest in reducing and focusing the role of the political process in order to maximize efficiency: a tax framework would do that. An environmental tax also raises the visibility of the costs associated with global warming abatement. This visibility should tend to limit the public's desire for environmental legislation because the cost would appear explicitly on utility bills.

An environmental tax framework also has a great deal to offer the environmental cause. First, the scale of the problem gives urgency to the issue of cost-effectiveness. At some point, the cost to society cannot be hidden, and that burden, coupled with limited results, could halt progress. More can be done if greenhouse emission reduction is done more efficiently. Second, the establishment of an environmental tax for global warming would create a framework that would serve the environmental cause well as the issue of global warming evolves. Currently, the weakest part of the proabatement argument is the uncertainty of the impact of global warming. Thus, the first round of policy actions could easily be a watered-down version of what might be necessary. (For example, the tax might be too low, too many emissions permits might be created, or too few controls might be mandated.) As scientific inquiry narrows the range of uncertainty, and assuming that the expected impact remains about the same (or increases), it is extremely likely continued modifications of policy will be needed. Responding to changes in scientific understanding under a tax regime would simply require an adjustment in the tax level. Adapting a control regime would be much more difficult.

Environmentalists should also be concerned about how different abatement policies dispose of property rights: those whose property rights are affected may resist needed change. Any control method, of course, creates winners and losers. As discussed above, command-and-control regimes rigidly embed such property rights in specific technology standards: a firm may pollute so long as its equipment has been approved. This outcome is potentially harmful.

As discussed, emissions-trading regimes also grant property rights in a troublesome fashion. Consider what might happen if greenhouse gas abatement policies were extended to gases other than carbon dioxide. If it is found that abatement for methane or other gases is relatively inexpensive, sensible policy might dictate stringent targets for these gases and more relaxed standards for carbon dioxide. But efficient carbon dioxide abaters might resist such change because it would reduce the value of their tradeable emissions permits. Efficient abaters of particular pollutants may have an interest in more stringent controls being applied to "their" pollutant. This problem exists for both tax and emissions-trading regimes with regard to the change in relative prices. However, trading systems must explicitly allocate emissions permits to governments or economic players. It is this allocation of property rights that may cause great difficulty over time. The environmental tax regime outlined here would distribute property rights in a less divisive fashion.

Supporters of carbon tax proposals do have an additional question to address. The failure of the Clinton administration's Btu tax, which was proposed in 1993 and quickly rejected by Congress, might suggest that the time for a tax approach has come and gone. The Btu tax had substantial flaws not shared by an emissions tax; in my view, these flaws led directly to its demise. First, the proposed Btu tax had no clear rationale. It was not clearly directed against environmentally damaging behavior because it taxed "clean" Btus as much as "dirty" Btus. It had little rationale as a pure revenue raiser because it was not based on the polluter's ability to pay. Second, it collapsed in part because of an overaccumulation of special preferences. The creators of the plan laid the groundwork for this collapse by setting the tax on coal Btus equivalent to that for much cleaner fuels and lower than that for oil, which is somewhat cleaner than coal. This preemptive capitulation to coal interests undercut a potential rationale for the tax and any strong stand against other preferential treatment. A carbon (or environmental) tax is clearly directed at a perceived harm and in principle admits no preferences. As such, a carbon tax need not be undercut by the flaws that plagued the Btu tax.

CONCLUSION

Global warming is a serious matter, and caution is appropriate in choosing initial policy responses for two reasons. First, many of the established policies have not been effective. Second, given the necessary scale of the potential policy responses to address global warming, taking the wrong path could be very expensive and could make the public and decision-makers skeptical of environmental initiatives. However, caution is not synonymous with inaction; rather, it counsels that the first action should be an investment in time and effort to establish a policy framework that can be effectively and efficiently implemented and that can facilitate co-gent political decision making under uncertainty. In particular, a policy framework should be flexible enough to respond to changing answers to the question, How much are people willing to pay to offset the probable effects of global warming?

Environmentalists may complain that such an emphasis can cause delay; it can. The establishment of the tax mechanism and debating and negotiating the tax level will be sources of delay. In addition, the tax level, at least initially, will likely be set lower than many think necessary. Decisionmakers might be less willing to vote for environmental initiatives when they must tie them directly to higher taxes rather than hide the costs for them in complicated technical proposals. Less comprehensive fixes could certainly be pushed through more quickly. But they would be much less effective and, in the end, might be much more expensive. The great problem of global warming requires better policies than have been used in the past.

NOTES

1. The World Bank has recently established a "Global Environment Facility," which does provide support, on a much lesser scale than contemplated here, for carbon dioxide reductions in developing countries.
2. The U.S. experience did demonstrate one positive political benefit from emissions trading. In congressional discussions, extra allowances for sulfur oxides became a currency to win the support of particular interest groups. Thus, additional allowances were created and distributed. Whether the political advantage of this type of logrolling can

carry over to an international arena is uncertain. The environmental consequences of "printing more money" (in this context, permits serve the same function as money) are adverse, of course.
3. See the chapter in this volume by Norberg-Bohm and Hart for a discussion of why technology transfer is difficult.
4. IEF support should not be confused with supporting development effort. That is, an incentive is provided to develop along less environmentally harmful lines than would otherwise occur. To the extent that alternative technology is either more costly or complex, funding could offset (up to the limit of the per unit subsidy) the incremental cost burden and facilitate technology transfer. I would suspect that a cleverly implemented support for this limited goal would absorb only a fraction of the tax revenues. (More analysis of this topic is needed.) If this is the case, excess tax revenues should be recycled within the economies that generate them, perhaps by reducing other taxes.

REFERENCES

Cambridge Energy Research Associates. 1992. "Special Report: Fueling the Race for the Presidency." Cambridge, MA: Cambridge Energy Research Associates.

Cohn, Linda R., and Roger R. Noll. 1991. *The Technology Pork Barrel.* Washington, DC: Brookings Institution.

Gaskins, Darius, and Bruce Stram. 1991. "A Meta Plan: A Policy Response to Global Warming." In *Opportunities for Collaborative Greenhouse Gas Research by the Electric Utility Industry.* Palo Alto, CA: Electric Power Research Institute, April.

Rosenberg, N., W. Easterling, P. Crosson, and J. Darmstadter. 1988. "Greenhouse Warming: Abatement and Adaptation." Proceedings of a workshop, Resources for the Future, Washington, DC, 14–15 June.

Tietenberg, T. H. 1986. "Uncommon Sense: The Program to Reform Pollution Control Policy." In *Regulatory Reform: What Actually Happened,* eds. Leonard W. Weiss and Michael W. Klass. Boston: Little, Brown.

U.S. Fuels Corporation. 1985. "Comprehensive Strategy Report," June. Fosters Report No. 1684. 11 August 1988.

Warrick, Bolin, and Doos Jager. 1986. *The Greenhouse Effect, Climatic Change and Ecosystems.* Chichester, England: Scientific Committee on Problems of the Environment.

8

The Economic Effects of a Carbon Tax

■ **DALE W. JORGENSON**
John F. Kennedy School of Government
Harvard University

■ **PETER J. WILCOXEN**
Department of Economics
University of Texas at Austin

A carbon tax, such as the one proposed by Bruce Stram in the previous chapter, would have effects throughout the economy. It would raise the price of energy, increase the cost of products produced by energy-intensive processes, reduce employment in energy sectors, increase employment elsewhere, generate tax revenue, and, finally, reduce carbon dioxide emissions.

In this chapter, we use a detailed model of the U.S. economy to calculate the magnitude of these and other effects. Our findings are roughly as follows. A carbon tax would raise fuel prices, particularly for coal. Coal demand would fall substantially, leading to a large drop in coal production. Oil and gas output would also decline, but by a much smaller percentage. Higher coal prices, in turn, would raise the cost of electricity. Consumers and firms would demand less electricity, which would slow productivity growth and capital formation. This, in turn, would tend to reduce GNP.

What actually happens to GNP, however, depends very strongly on how the revenue from the tax is used. A carbon tax large enough to have much effect on emissions would annually raise tens to hundreds of billions of dollars. If this revenue were used to reduce distortionary taxes elsewhere in the economy, the impact of the tax on GNP would be much smaller. In fact, we show that GNP would actually increase if the revenue were used to reduce taxes on capital.

The next section presents our model and uses it to show how a carbon tax large enough to stabilize carbon dioxide emissions at 1990 levels would affect the United States. We then consider a series of possible refinements on the basic tax: the further reduction of emissions, the different uses of the tax revenue, the alternative use of Btu or ad valorem taxes, and the distributional effect of the tax, and, finally, we explore whether a tax in OECD countries alone would be likely to improve the global environment. Appendix A of this chapter provides more detail about the model and the base case.

MODELING APPROACH

The results we present are based on a set of simulations we conducted using a detailed model of the U.S. economy that was designed specifically to examine the effects of energy and environmental policies. One feature of our approach that distinguishes it from many others is that we use a general equilibrium model. General equilibrium models are constructed by dividing the economy into a collection of interdependent sectors, which interact through markets for goods and services. (The behavior of each sector is represented by an appropriate submodel.) When a new policy, such as a carbon tax, is introduced, prices and wages adjust until demands and supplies are equated in every market and the economy reaches equilibrium.

Our model is composed of thirty-five producing sectors, a consumer sector, an investment sector, a government sector, and a foreign sector. Appendix A presents an overview of the model by describing the submodels used to represent each of these sectors. It also discusses our base case simulation.[1]

THE COST OF STABILIZING EMISSIONS

Somewhat surprisingly, the United States has had an extended period of stable carbon dioxide emissions once before: from 1972 to 1985. During that period, high oil prices reduced energy demand and lowered carbon dioxide emissions substantially. The relationship between oil prices and carbon emissions can be seen by a comparison of historical oil prices, shown in figure 8.1, with the history of U.S. carbon emissions, shown in figure 8.2. The large increases in oil prices in 1974 and 1979 led to drops in the trend rate of emissions growth. However, this reduction came at a very high price: the oil price shocks reduced U.S. GNP growth by 0.2 per-

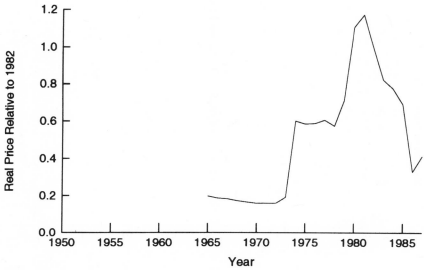

Figure 8.1. Real price of oil.

cent per year from 1974 to 1985.[2] The lesson from this episode is that a 0.2 percentage point of annual GNP growth is an upper bound on the cost of stabilizing U.S. carbon dioxide emissions.

The policy most often proposed for reducing carbon emissions is a carbon tax.[3] A carbon tax would be applied to fossil fuels used for combustion in proportion to the carbon dioxide the fuels emit when burned. From the standpoint of economic efficiency, a carbon tax is the ideal way to reduce carbon dioxide emissions because it is very close to a tax on the externality itself: if firms and individuals must pay to emit carbon dioxide, they will emit less. A carbon tax would stimulate users to substitute other inputs for fossil fuels and to substitute fuels with lower carbon content, such as natural gas, for high-carbon fuels such as coal.

Fossil fuels differ substantially in both price and the amount of carbon dioxide produced per Btu of energy (see table 8.1). One Btu from oil, for example, costs more than three times as much as a Btu from coal but produces only 80 percent of the carbon dioxide. The least carbon-intensive fuel is natural gas: one Btu of natural gas produces only about half as much carbon dioxide as does one Btu of coal. Together, the differences in price and carbon content mean that a carbon tax would produce very different percentage changes in the prices of the fuels. A $10 per ton tax on carbon would raise the price of coal by 29 percent, the price of oil by 6 percent, and the price of gas by almost 7 percent.

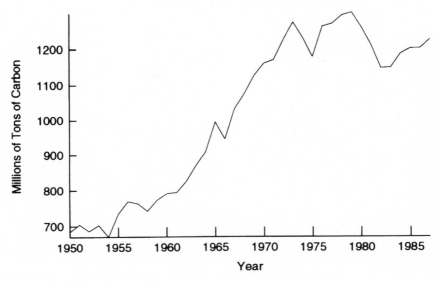

Figure 8.2. U.S. carbon emissions.

TABLE 8.1

	Fuel		
Characteristic	Coal (ton)	Oil (bbl)	Gas (mcf)
Million Btu per unit	21.94	5.80	1.03
Tons of carbon per unit	0.649	0.137	0.016
Carbon per million Btu			
Tons	0.030	0.024	0.016
Percentage relative to coal	100	80	54
Approximate price before tax ($)			
Per unit of fuel	22	21	2.4
Per million Btu	1.00	3.62	1.36
Tax equal to $10 per ton of carbon ($)			
Per unit of fuel	6.49	1.36	0.16
Per million Btu	0.30	0.24	0.16
Percentage increase per unit	29.50	6.48	6.67

Relative Carbon Content of Fossil Fuels (table caption)

The carbon tax policy that has been debated most widely would impose a tax large enough to limit emissions to 1990 rates. To measure the effect of such a policy on the United States, we constructed a simulation in which the carbon tax rate was allowed to vary from year to year but was always chosen to be exactly enough to hold U.S. carbon dioxide emissions at their 1990 value of 1,576 million tons.[4] We returned the revenue raised

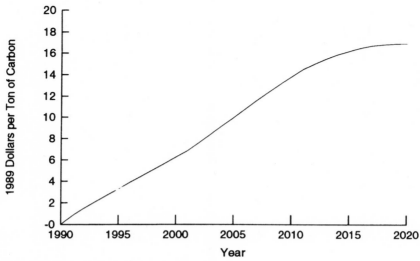

Figure 8.3. Carbon tax under stabilization.

by the tax to households as a lump sum rebate. Without the carbon tax, emissions would increase over time, so the tax grows gradually over the next few decades (see figure 8.3). By 2020 the U.S. population is likely to crest and emissions growth will begin to slow, reducing the rate of carbon tax growth.[5]

The tax would produce significant reductions in carbon emissions, as shown in figure 8.4. By 2020 emissions are 16 percent lower than they would have been without the tax. The tax also produces considerable revenue: $31 billion annually by 2020 (all dollar amounts are in 1990 prices).

The principal direct effect of the tax is to increase purchasers' prices of coal and crude oil. By 2020, for example, the tax reaches $22.71 per ton of carbon. As shown in table 8.2, this amounts to a tax of $14.75 per ton of coal, $3.10 per barrel of oil, or $0.37 per thousand cubic feet of gas. The tax would increase the prices of fuels but leave other prices relatively unaffected (see figure 8.5). The price of coal would rise by 47 percent, the price of electricity would rise by almost 7 percent (coal accounts for about 13 percent of the cost of electricity), and the price of crude oil would rise by around 4 percent. The prices of refined petroleum and natural gas utilities would rise because of the tax on the carbon content of oil and natural gas.

Changes in the relative prices for fuels would affect demands for each good and lead to changes in industry outputs (see figure 8.6). In the model, most sectors show only small changes in output. Coal mining is an

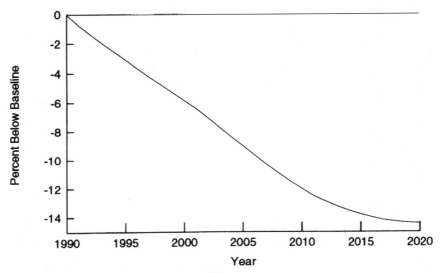

Figure 8.4. Carbon emissions under stabilization.

TABLE 8.2

Selected Results for the Stabilization Scenario, 2020

Variable	Unit	Value
Carbon emissions	% change	−16.12
Carbon tax	$ per ton	22.71
Tax on coal	$ per ton	14.75
Tax on oil	$ per bbl	3.10
Tax on gas	$ per mcf	0.37
Price of capital	% change	0.40
Capital stock	% change	−0.83
Tax revenue	$ (in billions)	31.41
Real GNP	% change	−0.55
Coal price	% change	46.99
Coal output	% change	−29.28
Electricity price	% change	6.60
Electricity output	% change	−6.17
Oil price	% change	4.45
Oil output	% change	−3.90

exception: its output falls by almost 30 percent. Coal is strongly affected for three reasons. First, coal emits more carbon dioxide than oil or natural gas per unit of energy produced. Thus, the absolute level of the tax per unit of energy content is higher on coal than on other fuels. Second, the tax is very large relative to the base case price of coal for purchasers: at the

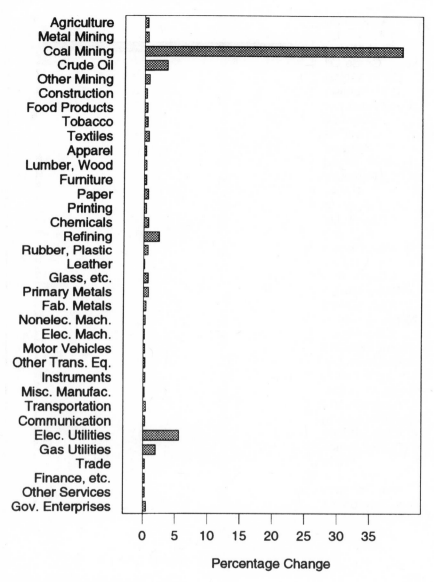

Figure 8.5. Effect on prices in 2020.

mine mouth, the tax would increase coal prices by around 50 percent. (In contrast, oil is far more expensive per unit of energy, so that in percentage terms its price is less affected by the tax. The price of crude oil rises only about 10 percent.) Third, the demand for coal is relatively elastic. Most coal is purchased by electric utilities, which can substitute other fuels for

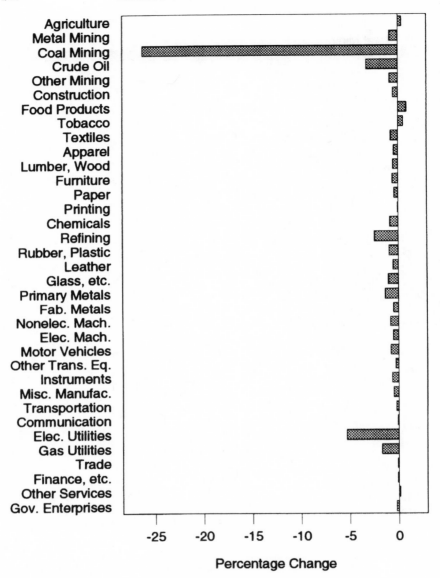

Figure 8.6. Effect on output in 2020.

coal when the price of it rises. Moreover, the demand for electricity itself is relatively elastic, so that when the price of electricity rises, demand for electricity (and hence demand for coal) falls substantially.

Outside the energy industry, the main result of the carbon tax would be to increase the prices of electricity, refined petroleum, and natural gas, each by a few percentage points. This rise in prices would have two effects. First, higher energy prices would mean that capital goods, which are produced through the use of energy, would become more expensive. Higher prices for capital goods would mean a slower rate of capital accumulation and lower GNP in the future. Second, higher energy prices would discourage technical change in industries in which technical change tends to increase energy intensity. Together, these two effects would cause the capital stock to drop by 0.7 percent and GNP to fall by 0.5 percent by 2020 (relative to the base case). Average annual GNP growth over the period from 1990 to 2020 would be 0.02 percentage points lower than in the base case. About half of this is due to slower productivity growth and half is due to reduced capital formation.

GREATER REDUCTIONS IN EMISSIONS

Even if emissions were stabilized at 1990 rates, atmospheric concentrations of carbon dioxide would continue to rise for decades. Thus, holding emissions at 1990 rates would not prevent further global warming. This observation has led many observers to call for much larger cuts in carbon dioxide emissions. To see how the economy would be affected by a more stringent carbon dioxide control policy, we constructed a second carbon tax scenario in which emissions were required to decrease 20 percent below 1990 levels by 2010. (Key results are shown in table 8.3; results from the stabilization simulation are also shown for comparison.)

Increasing the stringency of the policy increases its cost substantially. Moving from the stabilization scenario to the 20-percent reduction case doubles the effect of the policy: emissions fall by 32.90 percent instead of by 16.12 percent. However, this reduction is achieved at the cost of tripling both the carbon tax and the loss of output. The more stringent policy has a relatively larger effect on sectors other than coal mining. In particular, doubling the emissions reduction does not cause the reduction in coal output to double. Coal users, particularly electric utilities, find it increasingly difficult to substitute other fuels for coal. Thus, the larger reduction in carbon emissions requires a larger reduction in oil use. Under the 20-percent reduction case, the drop in oil use is three times what it was under stabilization.

We determined the economy's cost curve for a variety of emissions targets. The results are summarized in figure 8.7, which shows the reduction in U.S. GNP in 2020 as a function of the percentage by which emissions are below 1990 levels. The reduction of emissions by more than 20

TABLE 8.3

Selected Results for Two Emissions Targets

Variable	Unit	Reduce by 20%	Stabilize at 1990 levels
Carbon emissions	% change	−32.90	−16.12
Carbon tax	$ per ton	74.49	22.71
Price of capital	% change	1.10	0.40
Capital stock	% change	−2.35	−0.83
Tax revenue	$ (in billions)	82.52	31.41
Real GNP	% change	−1.71	−0.55
Coal price	% change	149.86	46.99
Coal output	% change	−55.03	−29.28
Electricity price	% change	19.46	6.60
Electricity output	% change	−16.43	−6.17
Oil price	% change	15.06	4.45
Oil output	% change	−12.35	−3.90

percent causes large losses in GNP. To put these numbers in perspective, stabilizing the atmospheric concentration of carbon dioxide, which would lead to an eventual stabilization of temperature, would require reducing emissions by 50 percent relative to 1990, a very costly policy.

USE OF CARBON TAX REVENUE

Any tax large enough to reduce carbon dioxide emissions significantly will raise an enormous amount of revenue. In the simulations above, the tax produces $30 to $80 billion a year. Precisely how this revenue is used will have a large effect on the overall economic cost of slowing global warming. In particular, if the revenue were used to reduce distortionary taxes elsewhere in the economy, or if it were used to lower government budget deficits, there would be large welfare gains, which would offset some or all of the welfare losses associated with the carbon tax itself.

To determine how large this welfare improvement might be, we constructed three simulations in which the revenue from a carbon tax was used to reduce different taxes. In each simulation, we imposed a carbon tax of $15 per ton in 1990, with the rate rising by 5 percent annually in subsequent years. In the first simulation, the revenue was returned to households by a lump sum rebate; in the second, the revenue was used to lower taxes on labor, such as social security taxes; and in the third, it was used to lower taxes on capital, such as corporate income taxes.

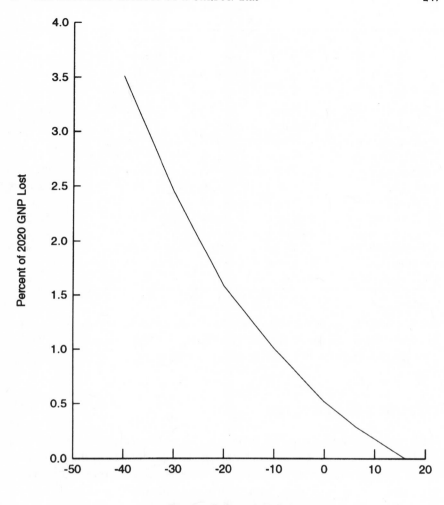

Percent Below 1990 Emissions

Figure 8.7. Cost of different emissions goals.

Our results show that the disposition of revenue from a carbon tax has a very significant effect on its overall impact on GNP (see table 8.4). In the lump sum case, output in 2020 drops by 1.70 percent relative to the base case. When the revenue is returned by lowering the tax on labor, the loss of GNP is less than half as much: only 0.69 percent. The improvement is due to an increase in employment brought about by the drop in the difference between before- and after-tax wages. If the revenue were returned as a reduction in taxes on capital, GNP would actually increase above its base case level by 1.10 percent. In this case, the gain is due to

TABLE 8.4

Selected Results for Revenue Experiments, 2020

| | | Revenue Policy | | |
| | | Lump sum | Labor | Capital |
Variable	Unit	rebate	rebate	rebate
Carbon emissions	% change	−32.24	−32.09	−31.65
Carbon tax	$ per ton	64.83	64.83	64.83
Price of capital	% change	0.97	−1.86	0.23
Capital stock	% change	−2.13	−1.36	1.89
Tax revenue	$ (in billions)	79.65	79.82	80.35
Real GNP	% change	−1.70	−0.69	1.10
Coal price	% change	143.49	140.57	142.06
Coal output	% change	−54.14	−54.19	−53.45
Electricity price	% change	18.57	15.97	16.99
Electricity output	% change	−15.93	−15.37	−14.66
Oil price	% change	14.20	12.28	14.55
Oil output	% change	−11.92	−11.54	−11.39

accelerated capital formation generated by an increase in the after-tax rate of return on investment. These results suggest that a carbon tax would provide an opportunity for significant tax reform.

OTHER TAX POLICIES

The concentration of costs in the coal sector raises the possibility that the coal lobby would be able to block passage of a carbon tax in the U.S. Congress. As a result, two alternative taxes are sometimes proposed: a tax on the energy content of fossil fuels (a Btu tax) and an ad valorem tax on fuel use. (An example of an ad valorem fuel tax that has often been proposed is an increased tax on gasoline.) Like a carbon tax, both of the other taxes would operate by raising the cost of fuels and hence inducing fuel users to substitute away from fuel use. This reduction in energy use is often proposed as a goal for its own sake. In a separate paper (Jorgenson and Wilcoxen 1993), we compared energy and ad valorem taxes to a carbon tax and found that although carbon taxes have the largest effect on coal mining, they have the smallest overall effect on the economy as a whole. Energy taxes were fairly similar to carbon taxes but had a slightly lower impact on coal mining (a drop in output of 25.0 percent instead of 26.3 percent) and a slightly higher overall cost (a drop in GNP at 2020 of 0.6 percent instead of 0.5 percent). In contrast, ad valorem taxes fell much

TABLE 8.5
Effects of Different Tax Instruments, 2020

Variable	Unit	Instrument		
		Carbon tax	Btu tax	Ad valorem tax
Carbon emissions	% change	−14.4	−14.4	−14.4
Carbon tax	$ per ton	16.96	—	—
Btu tax	$ per million Btu	—	0.47	—
Ad valorem tax	%	—	—	21.6
Tax on coal	$ per ton	11.01	10.21	—
Tax on oil	$ per bbl	2.31	2.70	—
Tax on gas	$ per mcf	0.28	0.48	—
Tax revenue	$ (in billions)	26	31	53
Capital stock	% change	−0.7	−0.8	−1.4
Real GNP	% change	−0.5	−0.6	−1.0
Price of coal	% change	39.9	37.2	26.1
Quantity of coal	% change	−26.3	−25.0	−19.5
Price of oil	% change	3.6	5.0	12.8

more lightly on coal mining (a 19.5-percent drop in output rather than a 26.3-percent drop) at the expense of having a much greater effect on the rest of the economy through higher oil prices (a 1.0-percent drop in GNP instead of a 0.5-percent drop; see table 8.5).

DISTRIBUTIONAL EFFECTS

Carbon taxes are sometimes opposed on the grounds that they are regressive. It is certainly true that a carbon tax could have widely varying effects across households. However, it is not clear that the tax would be significantly regressive.[6] Poterba (1991) estimated the impact of a $100 per ton carbon tax on U.S. households with different levels of total expenditure. He concluded that the impact of a carbon tax would be slightly regressive by this measure, falling more heavily on households having low total expenditures. Classifying households by income rather than by expenditure makes the tax appear slightly more regressive. DeWitt, Dowlatabadi, and Kopp (1991) conducted a study with more regional detail and found that there would be substantial differences in economic impacts across different geographic regions.[7] Both Poterba and DeWitt, Dowlatabadi, and Kopp came to the conclusion that nonenergy prices would also change, so that a general equilibrium approach is required to assess the full impact. This approach has been taken by Jorgenson, Slesnick, and

Wilcoxen (1992), who found the tax to be mildly regressive, although the size of the effect varied across different consumer groups.[8]

INTERNATIONAL ASPECTS

Unlike many environmental problems, carbon emissions are a global externality, which makes the implementation of a carbon tax difficult for several reasons. First, the tax would have to be levied by individual governments, some of which might not be willing to participate. Although most OECD nations have now agreed that some limit should be placed on carbon dioxide emissions, many developing nations have been reluctant to adopt any carbon dioxide policy that might reduce their economic growth. Schelling (1992) suggested that this resistance poses an insurmountable obstacle to a unanimous international policy. A likely outcome is that any global carbon dioxide policy would be incomplete: OECD nations would adopt the policy, while developing nations would not.

A tax with only partial international coverage could be vitiated by the movement of energy-intensive industries away from participating countries to other nations. In fact, Hoel (1991) has shown that if nonparticipating nations have inefficient energy technologies, it is theoretically possible for such a policy to result in a net increase in world carbon dioxide emissions. To date, however, only a modest amount of empirical research has been done on how an incomplete carbon policy would affect patterns of international trade. The principal study was conducted by Felder and Rutherford (1992), who found that the amount of redirected emissions could be considerable. An OECD carbon tax could reduce OECD oil demand enough to lower the world price of oil substantially. Lower world oil prices would lead, in the Felder-Rutherford model, to a large increase in oil demand by developing countries.

A second reason why the global nature of carbon dioxide emissions would make the implementation of a carbon tax difficult is that the point at which the tax is applied has important distributional effects. Whalley and Wigle (1990) noted that carbon taxes could be applied in several different ways, each achieving the same reduction in carbon dioxide emissions but having large differences in the distribution of costs.[9] A carbon tax large enough to reduce emissions substantially would raise an enormous amount of revenue. If the tax is applied at the point of production, it would be collected by the governments of producing countries. If the tax is applied to consumption, on the other hand, the revenue would flow to governments of consuming nations. Since the revenues are likely to be large, this is an important issue.

CONCLUSION

In this chapter, we examined the likely economic effects of a carbon tax. We find that a carbon tax will reduce U.S. GNP relative to its level in the absence of the tax, because a carbon tax will raise the price of energy. Moreover, the effects of the tax will be very similar to the effects of a tax placed solely on coal. Of all fossil fuels, coal is the least expensive per unit of energy and produces the most carbon dioxide when burned. Thus, a tax levied on carbon dioxide emissions will raise the cost of coal-based energy far more in percentage terms than the price of energy derived from oil or natural gas. In response to this price change, the demand for coal will fall substantially. The demands for oil and natural gas will also decline, but by much smaller percentages.

Almost all of the coal consumed in the United States is used to generate electric power. As the price of coal rises, electric utilities will convert some generating capacity to other fuels. However, substitution possibilities are fairly limited, particularly in the short run, so the tax will raise the price of electricity significantly. Consumers and firms will devote more effort to conserving energy by substituting other inputs for electricity, leading to a fall in electricity demand. Higher energy prices will lead to slower productivity growth, reduced capital formation, and a reallocation of labor to lower wage industries, all of which will cause GNP to be lower than it would have been in the absence of the tax.

The tax rate needed to achieve a fixed absolute emissions target, such as maintaining emissions at 1990 levels, will depend on how fast emissions grow in the absence of the tax. Baseline emissions growth, in turn, will depend on the rate of productivity growth, the rate of capital accumulation, the rate of growth of the labor force, any energy-saving biases in technical change, and the path of world oil prices. More rapid economic growth will generally lead to higher baseline emissions and will thus require higher tax rates if emissions are to be held at a fixed absolute level. Moreover, deeper absolute cuts in emissions will require sharply increasing tax rates.

A carbon tax large enough to have much effect on emissions would raise tens to hundreds of billions of dollars annually. Thus, it would provide an opportunity for significant tax reform, and this reform could soften the effect of the tax on GNP. If the revenue was used to reduce distortionary taxes elsewhere in the economy, the impact of the tax on GNP would be reduced. In fact, it is possible that GNP would actually increase if the revenue was used to reduce taxes on capital.

Finally, several additional observations need to be kept in mind regarding carbon taxes. First, the United States did, in fact, stabilize its carbon dioxide emissions during the 1970s and early 1980s when oil prices were very high. However, the cost in terms of lost GNP was much higher than if emissions had been controlled by a carbon tax. Costs would also be much higher if other inefficient instruments, such as a Btu or an ad valorem tax, were used instead of a carbon tax. Second, the distributional effect of a carbon tax will be regressive, but only very slightly. Third, if an international carbon tax were imposed in some countries but not in others, changes in trade patterns would shift carbon-intensive activities to countries in which they were not taxed, compromising the original policy to some extent.

APPENDIX A

OVERVIEW OF THE GENERAL EQUILIBRIUM MODEL AND BASE CASE

This appendix presents an overview of the general equilibrium model and base case that we used to estimate the effects of a carbon tax on the United States. The model includes thirty-five producing sectors, a consumer sector, an investment sector, a government sector, and a foreign sector. For more detail on the specification of the model or the base case simulation, see Jorgenson and Wilcoxen (1990).

PRODUCTION

In the model, production is disaggregated into the thirty-five industrial sectors listed in table A.1. Most of these industries match two-digit sectors in the Standard Industrial Classification. Each industry produces a primary product and may produce one or more secondary products. This level of industrial detail makes it possible to measure the effect of changes in tax policy on relatively narrow segments of the economy. Since most anthropogenic carbon dioxide emissions are generated by fossil fuel combustion, a disaggregated model is essential for capturing differences in the response of each sector to a carbon dioxide control policy.

The behavior of each industry is derived from an industry-specific nested cost function. At the highest level, the cost of each industry's output is assumed to be a transcendental logarithmic (translog) function

TABLE A.1

Industry Definitions

Number	Description
1	Agriculture, forestry, and fisheries
2	Metal mining
3	Coal mining
4	Crude petroleum and natural gas extraction
5	Nonmetallic mineral mining
6	Construction
7	Food and kindred products
8	Tobacco products
9	Textile mill products
10	Apparel and other textile products
11	Lumber and wood products
12	Furniture and fixtures
13	Paper and allied products
14	Printing and publishing
15	Chemicals and allied products
16	Petroleum refining
17	Rubber and plastic products
18	Leather and leather products
19	Stone, clay, and glass products
20	Primary metals
21	Fabricated metal products
22	Machinery, except electrical
23	Electrical machinery
24	Motor vehicles
25	Other transportation equipment
26	Instruments
27	Miscellaneous manufacturing
28	Transportation and warehousing
29	Communication
30	Electric utilities
31	Gas utilities
32	Trade
33	Finance, insurance, and real estate
34	Other services
35	Government enterprises

of the prices of capital services, labor, energy, and materials. The price of energy, in turn, is assumed to be a translog function of prices of coal, crude petroleum, refined petroleum, electricity, and natural gas, while the price of materials is a translog function of the prices of all other intermediate goods. Given this structure, we derived demand equations for capital services, labor, and intermediate inputs for each of the industries.

We estimated the parameters of each industry submodel econometrically through the use of a set of consistent interindustry transactions tables constructed for the purpose. The tables describe the U.S. economy for the period 1947 through 1985. (See Jorgenson and Wilcoxen, 1990, for details on how the data set was constructed.) Estimating the production parameters over a long time series ensures that each industry's response to changes in prices is consistent with historical evidence.

An unusual feature of our model is that productivity growth is determined within the model. Other models that have been used to study global warming—for example, Manne and Richels's (1992)—take productivity growth to be exogenous. In our model, the rate of productivity growth in each industry is determined endogenously as a function of input prices. In addition, each industry's productivity growth may shift it toward some inputs and away from others. Biased productivity growth is a common feature of historical data but is often ignored when production is modeled. By allowing for biased productivity growth, our model is able to capture the evolution of industry input patterns much more accurately.

CONSUMPTION

The model represents consumer behavior by assuming that households follow a three-stage optimization process. At the first stage, each household allocates full wealth (the sum of financial and human wealth, plus the imputed value of leisure time) across different periods. We formalized this decision by introducing a representative agent that maximizes an additive intertemporal utility function subject to an intertemporal budget constraint. The portion of full wealth allocated to each period is called "full consumption." At the second stage, households allocate each period's full consumption to goods and leisure in order to maximize an indirect utility function. This generates demands for leisure and goods as functions of prices and full consumption. The demand for leisure implicitly determines labor supply, while the difference between current income and consumption of goods implicitly determines savings.

The third stage of the household optimization problem is the allocation of total expenditure among capital services, labor services, and the thirty-five commodities. At this stage, we relaxed the representative consumer assumption in favor of the approach used by Jorgenson, Lau, and Stoker (1982) that derives separate systems of demand functions for households with different demographic characteristics. The model distinguishes among 1,344 household types according to demographic characteristics such as the number of household members and the geographic region in which the household is located. The spending patterns of each household type are derived from a hierarchial tier-structured indirect utility function. This generates household demands for individual commodities.

As with production, the parameters of the behavioral equations for all three stages of our consumer model are estimated econometrically. Our household model incorporates extensive time-series data on the price responsiveness of demand patterns by consumers and also makes use of detailed cross-section data on the effects of demographic characteristics on consumer behavior. In addition, an important feature of our approach is that we do not require that the pattern of household demands be independent of income. (Formally, we do not impose the restriction that the utility function be homothetic.) As total expenditure increases, spending patterns may change even in the absence of price changes. This method captures an important feature of cross-sectional expenditure data that is often ignored.

INVESTMENT AND CAPITAL FORMATION

The model has a single capital stock, which is in fixed total supply in the short run. This capital is perfectly malleable and can be reallocated among industries and between industries and final users at zero cost. Thus, the price of a unit of capital services is equal in every industry, and there is a single economywide rate of return on capital.

In the long run, the supply of capital is determined by investment. Our investment model is based on the assumption that investors have rational expectations and that arbitrage occurs until the present value of future capital services is equated to the purchase price of new investment goods. This equilibrium is achieved by adjustments in prices and the term structure of interest rates. New capital goods are produced from individual commodities according to a model identical to those for the industrial sectors, so that the price of new capital depends on commodity prices. We estimated the behavioral parameters for new capital goods production using final demand data for investment over the period 1947–1985. Thus, the model incorporates substitution among inputs in the composition of the capital.

GOVERNMENT AND FOREIGN TRADE

The two remaining parts of the model are the government and foreign sectors. To specify government behavior, we began by computing total government spending on goods and services. We started by assuming that tax rates would be fixed at current levels in the absence of changes in policy. We then applied these rates to taxable transactions in the economy to obtain total tax revenue. To this we added the capital income of government enterprises and non–tax receipts to obtain total government revenue. (The capital income of government enterprises is endogenous, while non–tax receipts are exogenous.) Next we assumed that the government budget deficit can be specified exogenously and added the deficit to total revenue to obtain total government spending. To arrive at government purchases of goods and services, we subtracted interest paid to holders of government bonds together with transfer payments to domestic and foreign recipients. We then allocated spending among commodity groups according to fixed shares constructed from historical data.

In modeling the foreign sector, we began by assuming that imports are imperfect substitutes for similar domestic commodities. The mix of goods purchased by households and firms reflects substitution between domestic and imported products. We estimated the price responsiveness of this mixture econometrically from historical data. In effect, each commodity is assigned a separate elasticity of substitution between domestic and imported goods. Since the prices of imports are given exogenously, intermediate and final demands implicitly determine the quantity of imports of each commodity.

Exports are determined by a set of isoelastic export demand equations, one for each commodity, that depend on foreign income (which we take to be exogenous) and the foreign prices of U.S. exports. Foreign prices are computed from domestic prices by adjusting for subsidies and the exchange rate. The demand elasticities in these equations are estimated from historical data. Without an elaborate model of international trade, it is impossible to determine both the current account balance and the exchange rate endogenously. We chose to make the exchange rate endogenous and the current account exogenous.

The Base Case

To assess the effect of a carbon tax, we first determined the future path of the U.S. economy in the absence of the tax. To construct such a scenario, which we call a "base case," we adopted a set of default assumptions about the time path of each exogenous variable in the absence of changes in government policy. Since savings and investment are determined by the expectations of households and investors, we specified the values of the model's exogenous variables far into the future. Through the period 1990–2050, we forecast values of the exogenous variables on the basis of their behavior in the sample period. After 2050 we assumed the variables remain constant at their 2050 values to allow the model to converge to a steady state by the year 2100.

Our projections for 1990–2050 were made as follows. First, all tax rates were set to their values in 1985, the last year in our sample period. Next, we assumed that foreign prices of imports in foreign currency remain constant in real terms at 1985 levels. We then projected a gradual decline in the government deficit through the year 2025, after which the nominal value of the government debt is maintained at a constant ratio to the value of the national product. Finally, we projected the current account deficit by allowing it to fall gradually to zero by the year 2000. After that, we projected a current account surplus sufficient to produce a stock of net claims on foreigners by the year 2050 equal to the same proportion of national wealth as in 1982. That is, we projected the United States to return to its 1982 position of being a net lender to the rest of the world.

Some of the most important exogenous variables are those associated with the growth of the U.S. population and corresponding changes in the economy's time endowment. We projected population by age, sex, and educational attainment through the year 2050 using demographic assumptions consistent with the Social Security Administration's projections. After 2050 we held population constant, which is roughly consistent with social security projections. In addition, we projected the educational composition of the population by holding the level of educational attainment constant beginning with the cohort reaching age 35 in the year 1985. We transformed our population projection into a projection of the time endowment by assuming that the pattern of relative wages across different types of labor remains as it was in 1985. Since capital formation is endogenous in our model, our projections of the time endowment effectively determine the size of the economy in the more distant future.

ACKNOWLEDGMENTS

This project has been supported by the U.S. EPA under contracts 68-W8-0113 and 68-W1-0009 and by the National Science Foundation under grants SES-90-11463 and SES-91-10231. We are solely responsible for the views expressed in this chapter.

NOTES

1. For more detail on the specification of the model or the base case simulation, see Jorgenson and Wilcoxen (1990).
2. For a more complete discussion, see Jorgenson and Wilcoxen (1993).

3. A carbon tax was first proposed by Nordhaus (1979).

4. A tax that varies from one year to the next in order to keep carbon emissions absolutely constant is a useful analytical device but is not a likely policy. The tax could not be adjusted quickly enough to keep emissions constant in every year.

5. Our population projection is based on forecasts made by the U.S. Social Security Administration in which population growth approaches zero early in the next century.

6. The distributional effects of a carbon tax have been examined by Poterba (1991), DeWitt, Dowlatabadi, and Kopp (1991), Jorgenson, Slesnick, and Wilcoxen (1992), and by Schillo et al. (1992).

7. DeWitt, Dowlatabadi, and Kopp (1991) used a detailed econometric model of U.S. household energy consumption to estimate the response of energy consumption patterns to the tax.

8. Jorgenson, Slesnick, and Wilcoxen (1992) used a detailed, econometrically estimated intertemporal general equilibrium model to measure the lifetime incidence of a carbon tax on consumers in different demographic groups.

9. Whalley and Wigle (1990) used a dynamic global general equilibrium model to assess the distributional effects of various carbon dioxide abatement policies.

REFERENCES

DeWitt, Diane E., Hadi Dowlatabadi, and Raymond J. Kopp. 1991. "Who Bears the Burden of Energy Taxes?" Discussion paper QE91-12. Resources for the Future, Washington, DC.

Felder, Stefan, and Thomas F. Rutherford. 1992. "Unilateral Reductions and Carbon Leakage: The Consequences of International Trade in Oil and Materials." May. Mimeo. The University of Colorado, Boulder.

Hoel, Michael. 1991. "Global Environment Problems: The Effects of Unilateral Actions Taken by One Country." *Journal of Environmental Economics and Management* 20, no. 1: 55–70.

Houghton, R. A., and G. M. Woodwell. 1989. "Global Climate Change." *Scientific American* 260, no. 4 (April): 36–47.

Jorgenson, Dale W., Laurence J. Lau, and Thomas M. Stoker. 1982. "The Transcendental Logarithmic Model of Aggregate Consumer Behavior." In *Advances in Econometrics*, vol. 1, eds. R. L. Basmann and G. Rhodes. Greenwich, CT: JAI Press, 97–238.

Jorgenson, Dale W., Daniel T. Slesnick, and Peter J. Wilcoxen. 1992. "Carbon Taxes and Economic Welfare." In *Brookings Papers on Economic Activity: Microeconomics Issue*, 393–431.

Jorgenson, Dale W., and Peter J. Wilcoxen. 1990. "Intertemporal General Equilibrium Modeling of U.S. Environmental Regulation." *Journal of Policy Modeling* 12, no. 4: 715–744.

Jorgenson, Dale W., and Peter J. Wilcoxen. 1991a. "Global Change, Energy Prices, and U.S. Economic Growth." *Structural Change and Economic Dynamics* 3, no. 1: 135–154.

Jorgenson, Dale W., and Peter J. Wilcoxen. 1993. "Reducing U.S. Carbon Dioxide Emissions: An Assessment of Different Instruments." *Journal of Policy Modeling* 15, no. 5–6 (October–December 1993): 491–520.

Jorgenson, Dale W., and Peter J. Wilcoxen. 1992. "Reducing U.S. Carbon Dioxide Emissions: The Cost of Different Goals." In *Advances in the Economics of Energy and Natural Resources*, vol. 7, ed. John R. Moroney. Greenwich, CT: JAI Press, 125–158.

Manne, Alan S., and Richard G. Richels. 1992. *Buying Greenhouse Insurance: The Economic Costs of CO_2 Emissions Limits*. Cambridge: MIT Press.

National Academy of Sciences. 1979. *Carbon Dioxide and Climate: A Scientific Assessment*. Washington, DC: National Academy of Sciences.

Nordhaus, William D. 1979. *The Efficient Use of Energy Resources*. New Haven: Yale University Press.

Nordhaus, William D. 1991. "To Slow or Not to Slow: The Economics of the Greenhouse Effect." *The Economic Journal* 101, no. 407: 920–937.

Poterba, J. M. 1991. "Tax Policy to Combat Global Warming: On Designing a Carbon Tax." In *Global Warming: Economic Policy Responses*, eds. R. Dornbusch and J. M. Poterba. Cambridge: MIT Press, 71–97.

Schelling, Thomas C. 1992. "Some Economics of Global Warming." *American Economic Review* 82, no. 1: 1–14.

Schillo, Bruce, Linda Giannarelli, David Kelly, Steve Swanson, and Peter J. Wilcoxen. 1992. "The Distributional Impacts of a Carbon Tax." February. Mimeo. Environmental Protection Agency, Washington, D.C.

Schneider, Stephen H. 1989. "The Greenhouse Effect: Science and Policy." *Science* 243, no. 4894 (February): 771–781.

Solow, Andrew R. 1990. "Is There a Global Warming Problem?" In *Global Warming: Economic Policy Responses*, eds. R. Dornbusch and J. M. Poterba. Cambridge: MIT Press.

U.S. Environmental Protection Agency. 1989. *Policy Options for Stabilizing Global Climate.* 3 vols. Draft report to Congress. February.

Whalley, John, and Randall Wigle. 1991. "Cutting CO_2 Emissions: The Effects of Alternative Policy Approaches." *The Energy Journal* 12, no. 1: 109–124.

9
Technological Cooperation: Lessons from Development Experience

■ **VICKI NORBERG-BOHM**
Department of Urban Planning
Massachusetts Institute of Technology

■ **DAVID HART**
John F. Kennedy School of Government
Harvard University

Any effort to mitigate global warming is bound to fail without the enthusiastic participation of developing countries. Although in the first half of the 1990s developing countries contributed only about 40 percent of greenhouse gas emissions, they will account for over 60 percent within a few decades. Even if industrialized countries stabilize their emissions at current levels, the risk of substantial global warming will remain high unless the growth of emissions in developing countries is slowed.[1]

Slowing the growth of greenhouse gas emissions will require consumers and enterprises to choose different technologies and governments to develop policies and programs to induce these changes. Developing countries have made it clear that they will not make these changes at the expense of economic growth, nor should they be expected to. While there are many technologies already in widespread use in industrialized countries that developing countries could use to substantially cut their greenhouse gas emissions while still meeting consumer and industrial demands, the acceleration of the purchase and use of these technologies will not come easily or without cost. Developing countries will need technological and financial support from industrialized countries if they are to adopt more environmentally benign technologies.

The form and dimensions of this support were among the most contentious issues in the process that led to the 1992 climate convention in

Rio de Janeiro. As a modest first step, the convention affirmed that industrialized countries will provide some assistance to developing countries to help them meet their treaty obligations. The specific arrangements to implement this pledge have been left unresolved.[2]

This chapter offers ideas that will be useful as the international community moves from these vague promises to the difficult steps of institutional and program design and implementation. Our thesis is that the experience that the international community has accrued in four decades of development programs provides important lessons about the promotion of technological change. Most broadly, experience shows that the social and political features of technological change are as important as its economic and physical features. Technological change will not occur unless consumers, enterprises, and governments change too. These complexities must be understood and integrated into international programs for mitigating the greenhouse effect. If they are, the developing countries can move toward a more environmentally sound development path, one in which technological innovations, institutional transformations, and new ways of thinking sustain one another.

The lessons we draw are broadly applicable to technological change in developing countries. However, programs for slowing the growth of greenhouse gas emissions will be sector specific. In this chapter, we focus on the energy sector. Energy production and use are responsible for a large and growing portion of developing countries' greenhouse gas emissions. Developing countries are expected to make huge investments in infrastructure and industry over the next several decades. If these investments push enterprises and consumers in developing countries along their present path of energy use, greenhouse gas emissions will skyrocket at a heavy cost to the global climate.

Fortunately, there are a variety of technological options that can reduce greenhouse gas emissions from energy production and use while providing the needed energy services. For example, countries with access to natural gas could substitute gas turbines for coal boilers in electricity generation and provide the same amount of power with half or less of the carbon dioxide emissions. Renewable energy sources are currently cost-competitive in some niche markets and provide electricity and heat with no carbon emissions. Carbon dioxide emissions could be reduced by improving electric power plant efficiencies and by reducing transmission and distribution losses. Energy-efficient technologies are available for a host of end uses, including lighting, appliances, building designs, office equipment, motors, industrial processes, and automobiles. U.S. Department of Energy analysts project that over the long term developing countries could use energy 30 to 60 percent more efficiently, thus substantially slowing the growth of their greenhouse gas emissions.[3]

Effective international cooperation programs will be needed if developing countries are to take advantage of these opportunities. In the next section, we discuss the principle of technological cooperation that should guide such programs. We then draw on this principle to present four specific lessons from development experience. These lessons pertain to consumers, enterprises, hardware, and political systems.[4] In each lesson, we first discuss the reasons why efforts to promote technological change often fail and then present strategies that should work better. The lessons are illustrated with contemporary programs and policies. The second section draws on these lessons to make recommendations for multilateral and bilateral technological cooperation programs. First, we identify three important themes from the lessons: institutional sustainability, public-private partnerships, and creative leadership. We then review and make recommendations for international and U.S. policy initiatives. We end by touching on the bottom line: the costs of technological change in developing countries' energy systems.

LESSONS FROM DEVELOPMENT EXPERIENCE

International development efforts have been based on two general conceptions of technology that have been tested and shown wanting over the last forty years. In the 1950s and 1960s, technology was viewed mainly as a freely available public good. Once political democracy, free markets, and an open international economic order were introduced, the thinking went, technology from industrialized countries could be easily appropriated by developing countries and would sustain their economic growth. When the sanguine hopes of this modernization theory were not realized, some observers came to see Western technology as perpetuating the dependency of developing countries and inhibiting their economic growth. New strategies of industrialization that controlled imports and restricted foreign investment were pursued, so that indigenous technological change could be cultivated. Dependency theory, however, did not prove any more satisfactory as a guide to technological practice than did modernization theory.[5]

We have learned that industrial development is neither an inevitable product of liberal institutions nor achievable by turning inward to national resources. Rather, successful industrialization in the late twentieth century appears to involve selective imitation of Western technology and institutions, leavened by technical and policy innovations that permit these imports to function well in a context far different from that in which they were created. Project-level experience has shown that technological change encompasses not only hardware, which comprises the machines and products that are typically associated with the word "technology," but also software, which includes skills and attitudes, and political and financial institutions. Technological change is a complex, multilevel process, depending on and affecting many features of individuals, groups, organizations, and societies. No single best, all-purpose policy package seems able to foster it.

The new vision of technology in development incorporates the idea that adopters of new technology should participate in technological choice and implementation. This hard-won knowledge grew from many development projects that have failed because they proceeded without local desire for, familiarity with, or ability to operate and service the hardware being promoted. Thus, there has been a change in language from "technology transfer"—in which the implied flow of expertise, knowledge, and hardware is unidirectional—to "technological cooperation"— in which technology is jointly constructed with domestic and imported parts, labor, and knowledge. Below we translate the abstract idea of technological cooperation into more specific lessons for programs aimed at slowing the growth of greenhouse gas emissions in developing countries.

LESSON 1: TECHNOLOGY ADOPTION

Consumers and enterprises tend to be risk-averse, ill-informed, and myopic. Energy price changes must be supported by policy and institutional innovation to overcome entrenched demand patterns.

There are many opportunities for consumers and producers in developing countries to use technologies that would reduce the risk of global climate change. The adoption of more energy-efficient products may be the largest. To bring these technologies into widespread use, governments must induce citizens to buy them. Experience has shown that altering demand patterns to meet policy goals is difficult. Even if the new product seems to be a better mousetrap according to standard economic analysis, potential adopters, particularly poor consumers, do not necessarily beat a path to the seller's door.

THE BARRIERS

High perceived costs, lack of information, and embedded habits impede better choices. Consumers are very sensitive to the initial cost of a product. In practice, new technologies have a very high discount rate, ensuring future savings far less than present savings. Compact fluorescent lamps, for instance, use about one-quarter as much electricity as incandescent lamps, and they last ten or more times longer. However, they cost about ten to fifteen times as much. The time needed to make up the difference in first cost depends on the cost of electricity and the usage of the lamp. Yet even in industrialized countries, where the payback period is typically only two years or less and the price of a compact fluorescent lamp is a small fraction of a consumer's income, market penetration has been slow. In developing countries, where the payback period is likely to be longer (because electricity is subsidized and lamps may be used less), this problem is magnified.[6]

Consumers may simply lack the capital to buy a product with a high first cost. Loans may not be available at reasonable rates of interest, or they may be viewed as risky. Consumers may also be so uncertain about their future earnings that they are reluctant to make the investment. In addition, they may be skeptical about savings that depend on factors outside their control, such as future energy prices.

Perhaps more important, consumer decisions are not based solely on prices and financing. Consumers are largely creatures of habit, and for good reason: they have little money or time to spare. Many of the technologies that could contribute substantially to greenhouse mitigation if widely adopted provide essential services, such as cooking, heating, and lighting. Consumers can ill afford an interruption in them. Consumers value reliability and ease of operation and often lack the information or ability to evaluate new or complex products. Furthermore, the cost of a large appliance—or even of a compact fluorescent lamp—may be a substantial fraction of an average family's income in a developing country. The failure or breakage of such a product could be a catastrophic loss of a capital investment. Consumers thus tend to be risk-averse, choosing products with which they are familiar.

Private and state-owned enterprises are also potential adopters of many technologies that could help to cut greenhouse gas emissions, and they demonstrate many of the same tendencies. Enterprises often are risk-averse and lack access to capital and information. As with consumers, they tend to be slow to adopt energy-saving technologies.

EFFECTIVE POLICY PACKAGES

Energy pricing is an important policy tool for the promotion of more efficient energy use. However, price alone will not stimulate optimal investment in energy conservation. Many energy-efficiency investments that would be cost-effective at current prices are currently not chosen in both the North and the South. Furthermore, energy in many developing countries is subsidized, making energy pricing part of a complex social contract. Raising prices, particularly to a level that would significantly slow the growth of greenhouse gas emissions, is politically difficult. Other policies, such as information provision, regulation, and the establishment of new institutions, must be used together with price reform.

The industrialized countries now have almost two decades of experience in trying to promote a more rapid diffusion of energy-efficient products. Some programs, such as technology demonstrations, energy audits, and appliance labeling, are aimed at providing potential adopters with enough information to overcome their reluctance to try new technologies. While better information may make some difference, it cannot change the short-term perspective of many consumers and enterprises. Regulations, such as appliance standards, building codes, automobile fuel efficiency standards, and emissions standards for industries, are one way of addressing this problem. For example, in the United States appliance standards have forced manufacturers to build more efficient products and to remove the least efficient designs from the market. Several developing countries have started similar energy conservation programs that have met with initial success.[7]

The most innovative policy efforts have focused on creating nongovernmental institutions that can take a long-term, systemwide perspective and make investments that are too costly for individuals but that can be justified from a social perspective. For example, many U.S. utilities have adopted integrated resource planning, a process that evaluates options for increasing supply and for reducing energy demand in determining how to meet electricity needs at the lowest cost to society and ratepayers. Integrated resource planners consider both new electricity-generating stations (supply-side technologies) and technologies that improve the efficiency of electricity use (demand-side technologies). Changes in regulation were necessary to create incentives for utilities to shift their perspectives and embrace this planning process.

The creation of institutions that can take a long-term approach is a difficult but important task. Utilities have played this role in the United States, but other types of organizations may be more suitable for those developing countries in which utilities are little more than patronage organizations. Local development banks, nongovernmental development organizations, and energy services companies are possibilities.

Multilateral development banks could provide the impetus for developing countries to create policies that foster such institutions. Development banks are the most important sources of funding for the energy sector in developing countries, and their influence extends to other investors and assistance agencies. However, they have only recently begun to incorporate energy efficiency requirements and integrated resource planning into their lending policy guidelines. Their traditional bias toward energy supply expansion stems, in part, from a lack of expertise about energy efficiency programs. Large energy supply projects also help them to meet large annual lending targets and are relatively easy to manage and monitor.[8] These barriers to greater lending for energy efficiency can be surmounted by the development of indigenous institutions (as described above) that can manage a large set of small energy efficiency investments. By cultivating and then channeling energy efficiency investments through these institutions, the banks would be able to meet their lending targets while helping energy efficiency to capture the prominent position that it deserves in the energy strategies of developing countries.

LESSON 2: ENTERPRISE TECHNOLOGICAL CAPABILITY

Enhanced technological capability is needed for enterprises in developing countries to contribute to greenhouse mitigation. To gain this capability, enterprises must invest in linkages and learning, investments which can be leveraged by international organizations and industrialized countries.

Enterprises make more complicated choices about technology than consumers; they not only adopt and use technologies but may also create and manufacture them. In making these decisions, however, all enterprises do not perform equally well, even given the same environment (such as market conditions for products and labor, and national and international policies). The policies and structures of each enterprise influence whether it attains command of state-of-the-art technologies.[9] Enterprises, both state-owned and private, that have invested human and monetary resources in developing technological capability—that is, routines and management strategies that foster awareness, learning, and innovation—have achieved commercial success. Enhanced technological capability will be needed for enterprises to take advantage of opportunities to get on a low emissions path. International cooperation should be aimed at aiding enterprises in developing countries to gain technological capability.

KEY INVESTMENTS

An enterprise's technological capability is embodied in its plant and equipment, in the skills of its employees, and in its management expertise. Capital purchases and formal training of workers and managers are important and well-known ways of developing technological capability. Development experience suggests that enterprises can benefit from other equally important, but often overlooked, investments in their technological capability.

The first is through technological gatekeeping, which tracks and takes advantage of developments made by others. An enterprise's technological gatekeepers not only keep abreast of the technical literature but also form links with other enterprises, professional and trade organizations, and research institutions. Good gatekeeping stretches the limited resources that enterprises have for technological development. In Mexico, for example, the Instituto de Investigaciones Eléctricas assists the national electric utility in the gatekeeping function through information exchange agreements with electric power research institutes in other countries, frequent interchanges with foreign experts, attendance at interna- tional conferences and workshops, a library that receives a broad range of technical publications, and a service that provides professionals within the utility a monthly summary of recent publications relevant to their specific areas of interest. The benefits that the national electric utility has received from this gatekeeping activity include improved specifications in contracts for high-technology equipment and assistance in problem solving in areas such as efficient operation of power plants, design and operation of the transmission and distribution system, and energy efficiency planning.[10]

Second, technological capability can be gained through a process known as "learning-by-doing." Access to process specifications, blueprints, licenses, or machinery does not ensure the ability to make a product because much technological information is tacit and cannot be communicated formally.[11] Such knowledge must be gained through operational experience in which employees focus on the systematic incorporation of new techniques. It is this systematic effort, including an active interest in both managers and employees obtaining feedback within the enterprise and a willingness to experiment with revised configurations and routines, that distinguishes learning-by-doing from merely doing.[12]

Enterprises that make an effort to learn as they do gain the ability to make incremental innovations. This ability is particularly significant in the energy sector because incremental technical and organizational changes typically lead to energy savings much greater than those available through simple housekeeping measures.[13] The re-engineering and re-configuration needed cannot be bought ready-made from suppliers; enterprises must be able to generate change internally. For this reason, policies designed to improve industrial energy efficiency must focus broadly on developing technological capability, and not narrowly on supporting purchases of efficient hardware.

Little about an enterprise's acquisition of technological capability is spontaneous. The enterprise must commit significant human and financial resources to devising and using new routines and strategies.[14] No enterprise can do so (and thus build up technological capability) unless its leadership's vision has a long time horizon. Indeed, investments in technological capability often must be actively championed by managers, who must defend them against competing needs for capital. Incentive structures that reward individual initiative and risk-taking—in other words, a culture of entrepreneurship—are common features of enterprises in developing countries that have successfully raised their technical standards. For example, the deep commitment by the management of USIMINAS, a Brazilian steel maker, to building technological capability has paid large dividends; in one instance, the enterprise was able to almost double its production over a period of six years through training, better materials and handling, process control, and quality management, without having to add new equipment.[15]

INTERNATIONAL SUPPORT: FROM INFORMATION EXCHANGE TO JOINT PROJECTS

Many international organizations have recognized the need to invest in the technological capability of enterprises in developing countries. But because these enterprises are diverse and vary widely in their technological capabilities, international programs are just beginning to address the full range of these enterprises' needs. We support more extensive and inventive efforts on this front.

Several efforts are under way to help share the costs of technological gatekeeping. Western governments, the UNEP, and international NGOs are establishing databases of the latest environmentally sound and energy-saving technologies available in the North.[16] International NGOs are also funding feasibility studies that compare alternative energy technologies. These studies are not confined to single decisions, such as whether to use cogeneration at a new facility, but cover system planning as well. For the most technologically capable enterprises in developing countries, facilitated access to technical information (and the provision of licenses for the use of new technologies when necessary) is likely to be an adequate mode of cooperation to achieve satisfactory greenhouse gas emissions levels at their plants.

However, many enterprises in developing countries are not highly capable, and access to information alone would be inadequate for them to contribute to greenhouse mitigation. To help these enterprises meet greenhouse gas emissions goals, the international community will need to support capability-building investments that go beyond gatekeeping in such areas as adaptive research and development, involvement in professional and industrial networks, expansion of maintenance facilities, training, and managerial assistance. The technology transfer provisions of the Montreal Protocol provide a useful precedent in this regard. For example, the multilateral fund established by the treaty will support training, research, and network building for the implementation of strategies to eliminate CFC use in developing countries.[17]

To increase the likelihood of success, international support for building the technological capability of enterprises in developing countries should be conditioned on investments in managerial and technological innovation by enterprises receiving assistance. These partnerships between international and national funders and enterprises in developing countries could be strengthened by the involvement of enterprises in industrialized countries that have the needed technological capability. Interfirm alliances are expanding across international borders within the industrialized world, but the extent to which they will involve developing countries remains doubtful.[18] While some enterprises in developing countries may be attractive partners because they can provide access to national markets that are otherwise closed, many appear to have little to offer in exchange for hardware, training, and management advice. Inventive funders may be able to catalyze such alliances by providing for incentives to overcome this barrier. The goal of this type of international technological cooperation would be to facilitate the development of enterprises' technological capabilities above and beyond what agreements among firms alone might achieve. To achieve this goal, such programs need to involve people from developing countries in all steps of these projects, from assessment through operation.

Joint demonstration projects for new technologies represent a particularly interesting type of North-South enterprise alliance. An example of this type of project is under way in Brazil, where a consortium of twelve companies, including private and public Brazilian enterprises and multinational firms, is developing a biomass gasifier/gas turbine power plant designed to use wood chips as fuel. The consortium was created by the joint entrepreneurship of foundations, industry associations, and governmental entities. While it is not yet commercially competitive, this technology holds great near-term promise as a means of competitively meeting the energy needs of developing countries with no greenhouse gas emissions.[19] The World Bank's Global Environment Facility has recently authorized $7 million for the design phase and will pay about one-third of the $75–80 million needed to construct and initially operate the plant.[20]

LESSON 3: COMPATIBLE TECHNOLOGY

Hardware must be compatible with the physical and institutional infrastructure.

The global economy is becoming more integrated, but factories are not the same the world over, nor is the world one big undifferentiated marketplace. The choice of technology and its evolution remain governed by local conditions—physical, cultural, and economic. The existence of a particular technological solution in the North does not mean that it is applicable or immediately usable in the South.

HARDWARE ADAPTATION AND LOCAL CONDITIONS

Anytime hardware is taken into a new environment, some innovation will be required to adapt it to that environment. This maxim is particularly true for products and equipment transferred from industrialized countries to developing countries, as they are often developed for use in quite different physical and institutional conditions. The Indian power equipment industry, for example, has adapted boilers originally developed in industrialized countries for low-ash coal to operate efficiently with the high-ash coal indigenous to India. Similarly, ballasts for compact fluorescent lamps must be redesigned to function effectively under the large frequency and voltage fluctuations that characterize most developing countries' electricity systems. Without such adaptation, the lamps burn out prematurely, reducing or eliminating the life cycle cost savings that are the main reason for choosing them.[21]

Products may require adaptation to the social environment as well as to the physical environment. For example, high-efficiency refrigerators in the North are often loaded with luxury features because the same consumers who value energy efficiency also value these sorts of luxuries, and can afford to pay for them. If such appliances are to penetrate markets in developing countries, simple energy-efficient models that more customers can afford must be developed.

The adaptation of technology that will meet local preferences and satisfy local needs and that will operate well in the local environment has often been summed up in the term "appropriate technology." By this definition, the goal of appropriate technology is unarguably worthy. However, "appropriate" may be a limiting concept if it assumes that societies are static. Appropriate, for instance, should not mean that labor is assumed to be perpetually cheap and uneducated. Local conditions may themselves be adapted to exploit a technological opportunity. Thus, while an unsuccessful attempt to introduce a new product may be blamed on inappropriate hardware, the failure may also be caused by improper training, inadequate complementary infrastructure, or poor institutional development.[22]

For example, the failure to diffuse biomass gasifiers for the fueling of irrigation pumps in the Philippines demonstrates how deficiencies in software can disrupt the dissemination of new technologies. The program's initial goal in 1981 was to replace over 1,000 diesel-powered pumps with charcoal gasifiers. Only 319 units were actually installed, however, and by 1987 only about 1 percent of these were in use. Poor maintenance, caused by inadequate user training and a centralized and poorly funded service and spare parts system, led to a high rate of equipment failure. Charcoal was in short supply and too expensive. Infrastructural weaknesses were compounded by institutional shortcomings. The agency in charge of the program lacked resources and enthusiasm, but because there was strong political pressure to move quickly, gasifiers were installed without sufficient testing. Finally, many farmers and operators of irrigation systems did not believe they would benefit from the new hardware.[23] The gasifiers themselves did not require adaptation so much as the system that supported them. Programs aimed at technological cooperation for greenhouse mitigation must be careful to attend both to adapting the hardware to fit local conditions and to modifying the conditions in which the hardware will be used.

NORTH-SOUTH PARTNERSHIPS TO ADAPT TECHNOLOGY

Our proposals for increasing the technological capability of enterprises are also relevant to the development of well-adapted hardware. Joint product development between enterprises in industrialized and developing countries, for instance, can ensure that local conditions are considered in the design process. To succeed in the dual tasks of adapting technology and creating new technological capability, such programs must include the involvement of local personnel and adequate investments in complementary assets, such as physical infrastructure and the capability to operate and maintain new technologies.

Developing countries often have trouble securing financial support to adapt technology. Development banks usually fund commercially viable technologies and avoid technology development. Western firms often have no incentive to engage in research for hardware adaptation or to develop technologies specifically for the markets of developing countries. As illustrated by the Brazilian gasifier project described above, the international community could play a constructive role through policy-assisted, market-oriented measures that facilitate the adaptation of technologies that mitigate greenhouse gases to the environments of developing countries and that support the creation of complementary infrastructure and technological capability.[24]

LESSON 4: POLICY ENVIRONMENT

Technological change is a political process in which social conflicts must be managed and contradictory policies must be reconciled. International programs for technological cooperation can foster domestic coalitions that support change and can build governing capacities to implement it.

The domestic political context in which technological change is pursued is particularly important, for it shapes the incentives for organizational and individual behavior, governs the human and financial resources on which enterprises can draw, and controls the relationship of the domestic economy to the global economy. Greenhouse mitigation strategies touch on so many aspects of economic life that they must be considered as a part of each nation's overall development strategy, which is most often an object of vast and deep political struggle.

OVERCOMING DOMESTIC POLITICAL CHALLENGES

Any existing technological system has stakeholders that want it to continue; as a system grows, its stakeholders grow with it and the system acquires momentum. Stakeholders include the enterprises that build and run the system and the consumers who use it, such as power customers and utilities in the case of electric systems. Stakeholders may also include the system's employees and their organizations, regulatory or promotional agencies whose budget and power are tied to the system's advancement, and even political parties or politicians who find either symbolic or economic gratification in the system. The state itself may rely on continued extension of the system; electric power, for example, integrates previously remote regions into the national economy and social life. Stakeholders will seek to defend the investments they have made in particular technologies.

Any political commitment to changing the course of a nation's technological development, such as that which might be implied in an international greenhouse treaty, will not be sustainable unless stakeholders in existing systems are addressed. Stakeholders that anticipate losses from change, such as utilities in a country in which the government desires to shift from electricity supply expansion to electricity demand management, are likely to protect themselves through the political process: through obstructing, delaying, regulating, and legislating. If they are powerful enough to block essential steps toward change, some way to give them a stake in change—or at least a stake in not preventing change—must be devised. A new coalition must be constructed around the prospective gains to be made from technological change. In the case of utilities under a demand management scheme, for example, one approach is to allow utilities to profit from the sale of demand-side technologies. The coalition-building process is inevitably messy. Not all stakeholders are easily identifiable in advance, nor can their political power be measured in the abstract. As at the enterprise level, the involvement of political entrepreneurs or technology champions will be pivotal.

International efforts to promote technologies that mitigate greenhouse gases must be sensitive to the political and social pressures that the leaders of developing countries will confront. At the least, strong majorities in developing countries will be expecting to expand their access to energy services. Likewise, most societies place a high priority on the growth of domestic manufacturing firms and electric utilities, both of which are seen as driving economic development. Thus, the political leaders of developing countries will be most attracted to greenhouse gas mitigation initiatives that will enhance their industrial and technological base. It is certain that developing countries will bargain hard in any global climate change agreement (as they did in the 1990 negotiations on technology transfer under the Montreal Protocol) to ensure that existing enterprises are not shrunk or made obsolete.[25]

Even if it would be less expensive in the short run to import more environmentally benign products rather than to try to manufacture them domestically, such a strategy may encounter significant domestic resistance in some developing countries. Strategies that support domestic manufacturing may be worth their initial higher costs because the political support such a strategy garners can substantially lower the risk of failure. In some cases, domestic enterprises may be able to shift production from old to new lines, for instance, from incandescent to fluorescent lamps. In other cases, new enterprises may enter an existing market with a new technology, as in the case of computerized lighting controls. Without such prospects, if a new product or its components must be imported or if the benefits of its sale are perceived to flow abroad, nationalism may become an effective opposition strategy. Furthermore, the extension of indigenous manufacturing capabilities may be the basis for future innovation, so that the long-term costs of greenhouse mitigation are lowered.

BUILDING STATE CAPACITY

The capacity of the state to construct a coherent policy environment is another important variable in the success of efforts to pursue a low greenhouse gas energy path. Governments will be most effective in facilitating technological change if they intervene simultaneously to promote both demand for and supply of new products. Policies in such diverse areas as economic development, education, science and technology, foreign affairs, and trade can have as much impact on building these markets as policies aimed directly at energy and the environment.

For instance, price subsidies, which are common in the energy markets of developing countries, may make investments in energy efficiency less attractive.[26] Similarly, tax schedules may undermine the economic viability of otherwise attractive projects. Enterprises will not adopt a new technology, no matter how much they are exhorted to do so by one ministry, if another ministry is expected to sap any potential gains. Governments can also have a profound influence on technological change, for good or ill, through protectionist policies, such as foreign exchange controls for purchases of technology and local content requirements. These policies are not easy to use well (though they have been powerful forces in the development of technological capability in Japan and South Korea in the past several decades) because they must be carefully calibrated to changing national and international circumstances.

The experience of the Bombay Efficient Lighting Large-Scale Experiment (BELLE), a project aimed at demonstrating the feasibility of broad residential use of compact fluorescent lamps, demonstrates some of the political complexities of technological change. In addition to overcoming the cultural gaps among the participating organizations—a utility, a lighting manufacturer, a social research institute, and U.S. funders and consultants—BELLE needed coordinated reform of utility regulation, tax, trade, and building-inspection policies across local, state, and national levels of government to succeed. BELLE faced consumer resistance, utilities with weak incentives to support energy efficiency, manufacturers that had little desire to innovate, and government bureaucracies that fiercely defended their turf. The government, for instance, was unwilling to lower its high import duties to allow the first shipment of compact fluorescent lamps to enter the country for demonstration purposes, even though indigenous manufacturing was planned for the future. After surmounting one barrier after another, the project ultimately stalled because it was unable to get access to a source of foreign exchange. It is ironic that although foreign donors were prepared to put up the money, its disbursal was controlled by domestic firms and research institutions whose interests conflicted with those of the project members.[27]

Effective policy making, coordination, and enforcement may put demands on the governments of developing countries that are beyond their present capacities. The state (or some legitimate designee, such as a regulated utility or a public-private consortium) may need to evaluate domestic enterprises, regulate trade, develop an industrial extension arm, facilitate training, or perform a wide variety of other tasks as part of the technological cooperation process. Training and technical assistance programs for government agencies may make the difference between success and failure in implementation. The international community has an interest in building national institutional capacities to permit governments in developing countries to undertake ambitious greenhouse response strategies.

As with enterprises, government agencies must learn by doing. Assistance programs will function best if linked to pilot projects. Only the actual implementation of policies and programs will give governments the experience they need to identify and overcome specific barriers, which can vary widely across countries, to planning and implementing reduction strategies for greenhouse gas emissions.

Several proposals are now circulating for the establishment of international networks of environmental research and energy efficiency institutes.[28] These networks could play an important role in technological cooperation efforts. However, the organizers of these efforts would do well to learn from the experience of the Consultative Group on International Agricultural Research (CGIAR). CGIAR initially encompassed only research institutions that concentrated on plant breeding, and it achieved striking success in raising the yields of staple grains like wheat and rice. However, increased grain production did not necessarily improve the income or position of poor farmers; the green revolution that the new hybrids spawned was sharply criticized for increasing the capital costs of farming by requiring the purchase of seeds, fertilizers, and pesticides. In the wake of this critique, some of the centers began to evaluate farming systems, including social relationships (such as land-holding patterns) and government policies (such as price subsidies), as well as continuing research on the crops themselves. New centers were added to CGIAR to conduct national and international policy analysis and to support national institution building in agricultural research and extension. Comparable organizations could be built into the proposed energy and environmental research networks from the beginning.[29]

IMPLICATIONS FOR THE INTERNATIONAL COMMUNITY, THE UNITED STATES, AND THE COST OF GREENHOUSE GAS MITIGATION

The four lessons that we have put forward have led us to recommend a diverse set of technological cooperation programs to support progress toward a global economy less likely to cause climate change. International efforts could help to reduce consumers' aversion to energy-efficient products, develop enterprises' technological capability, support the mutual adaptation of technology and local conditions, and cultivate a "greener" political environment in developing countries. Of course, the mix of programs must be tailored to suit each nation, and the most significant challenges vary widely. However, there are several common elements in our discussion that deserve to be highlighted and that have broad implications for international and U.S. policies and politics. We review these implications in the sections that follow, which also address the implications for financing technological change in the energy systems of developing countries, a topic we turn to in closing.

CONSIDERATIONS FOR PROGRAM DESIGN

INSTITUTIONAL SUSTAINABILITY

A recurring theme in this chapter is the difficulty of changing technological trajectories. The evolution of technology is closely linked to ways of thinking and patterns of power. To overcome entrenched interests, credible response strategies must not only make sense environmentally and economically—reducing emissions at an acceptable price—they must also make sense politically and organizationally. At the most fundamental level, any effort to lower greenhouse gas emissions must be integrated into national development planning. The organizations implementing a greenhouse policy must have adequate political clout and institutional capacity if these efforts are to be sustained and successful. Institutions such as enterprises, state agencies, and NGOs that can effectively challenge existing technological systems often do not exist in developing countries. Thus, the building of institutional capacities must be a central component of technological cooperation.

PUBLIC-PRIVATE PARTNERSHIPS

To have a substantial effect on greenhouse gas emissions, national governments must influence a large and varied group of actors, especially in the private sector. Their policies should structure market incentives so that consumers and enterprises will choose more environmentally sound products and techniques. Yet, in many cases, even with appropriate market signals, the risks and costs of creating or adopting socially desirable new technologies will be too high for the private sector to bear. Many programs that could address such market failures can best be organized by joint public-private endeavors. In programs as diverse as technology assessments, technology demonstrations, and new product development, innovative partnerships may help to ensure that private sector actors provide valuable inputs and have a stake in successful outcomes.

For example, a program funded by the U.S. Administration for International Development, the Program for the Acceleration of Commercial Energy Research (PACER) in India, is building institutional alliances between indigenous and U.S. energy technology manufacturers, research and development organizations, and end users by providing loans of up to $3 million for research and development and planning projects, but will cover no more than 50 percent of the total project cost. PACER is housed in a development bank, which acts as a venture capitalist to broker the new relationships. Proposals are screened by outside experts, who are overseen by a council consisting of senior representatives of public and private sector organizations.[30] By directing energy-related research toward commercial product development, PACER may be able to accelerate technological change in the Indian energy sector.[31]

CREATIVE LEADERSHIP

Energy systems encompass attitudes, habits, and organizations as well as hardware. They also exert inertia that can only be redirected through extended willful effort. This redirection requires leadership at many levels. Within enterprises, champions of innovation will have to overcome ingrained routines. At the project level, entrepreneurial public managers will need to convince recalcitrant actors to reconsider their own best interests. In these varied efforts for technological cooperation, creative leaders will play the important role of brokers, or bridging agents, who cross organizational and sectoral boundaries (for example, by providing information and structuring deals).[32]

At the policy level, national leaders must build winning coalitions for change in the face of many other pressing problems. Policies to slow the growth of greenhouse gas emissions are unlikely to move forward unless they simultaneously fulfill other priorities high on national development agendas. Such policies may shake up stable political coalitions or be viewed as serving the interests of industrialized countries. Visible, aggressive action on behalf of technological cooperation for greenhouse mitigation may be a risky proposition for political leaders in developing countries. Both multilateral and bilateral policies may assist in fostering and supporting these leaders.

CURRENT PROGRAMS AND FUTURE PROSPECTS

Negotiations on international technological cooperation for greenhouse mitigation have been largely devoted to the financial resources that would be needed and the hardware that would be made available if specific targets for emissions reductions are set. Other, complementary tasks will have an equally important effect on the success of any agreement. The international community must design initiatives and institutions to provide educational, technical, and managerial assistance to support actual implementation of technologies that emit low levels of greenhouse gases in developing countries.

We see promising signs on the international front. Some international organizations are already moving toward providing information about new products and production techniques. Potentially helpful additional roles in this arena include organizations participating in testing, publicizing, and serving as independent authorities on the performance of new technologies. Others in the international community are providing funds to promote indigenous technological capability through formal training and project feasibility studies and by funding developing countries' participation in international meetings. The international community might consider working harder to broker North-South partnerships between enterprises and to ensure that partnerships supported by international funding result in increased technological capability for the Southern participants. Technological assessments and country studies are also moving ahead. Support and assistance are needed to build national policy making, outreach, and monitoring capabilities as well.[33]

The United States can play a constructive role by supporting multilateral technological cooperation. The United States has also taken some promising environment and development initiatives on its own that deserve to be expanded further. One set of programs focuses on training, energy sector planning, institutional development, and project feasibility assessment. Although these technical assistance programs are too small to make capital investments, they may help to leverage technological change on a large scale by changing the attitudes and intellectual resources of program participants from developing countries. The United States could further leverage its investments in technological capability and project development by using its influence with development banks to refocus their energy sector loans to bolster strategies for developing countries that are environmentally as well as economically sound.[34]

Another set of programs promotes U.S. exports of renewable energy and energy efficiency technologies. These include trade missions, trade shows, and computerized data systems.[35] Such projects achieve some of the same goals as technical assistance programs by providing information and by funding feasibility assessments. Project financing is available through the U.S. Export-Import Bank and the Overseas Private Investment

Corporation.[36] Export promotion can provide a positive-sum approach to technological cooperation by simultaneously increasing U.S. exports, enhancing environmental quality, and improving technological capability in developing countries. Export promotion deserves a role in U.S. policy. However, it is not broad enough to form the foundation of a U.S. strategy for technological cooperation intended to slow the growth of greenhouse gas emissions, as some have suggested. Export promotion alone will not stimulate the broad range of activities needed to effectively promote low–greenhouse gas energy paths in developing countries.

Export promotion also raises the knotty issue of tied aid, that is, aid that requires recipient countries to purchase U.S. goods and services. To the extent that tied aid hinders adequate participation by individuals and enterprises from developing countries, it is likely to slow the development of technological capability and thus will undermine efforts to promote technological change. On the other hand, some tied aid may be necessary to build support for foreign assistance in the United States. Tied aid may also have the side benefit of building partnerships between enterprises and individuals in the United States and developing countries. The United States has traditionally exercised well-justified caution by tying its aid less than most other donors. Although tied aid programs may be beneficial in moderation, they should be evaluated to make certain that they promote the goal of developing indigenous capabilities and that they are not simply a payment to U.S. industry.

In the long run, the stimulation of innovation for the U.S. market is likely to be a more successful strategy for improving U.S. exports of environmentally friendly technology than tied aid. The best way to foster innovation is to enact a more aggressive domestic reduction strategy for greenhouse gas emissions. Such a strategy would bolster the incentives provided to U.S. companies to develop expertise and products that might be worthy of global diffusion. This strategy is particularly important to the small companies that constitute most of the domestic energy efficiency and renewable energy industries. The United States is currently competitive in some low–greenhouse gas energy technologies, but this situation could rapidly change. Japan, Germany, and other industrial competitors of the United States have already announced greenhouse mitigation strategies and have embarked on major new government-sponsored research and development initiatives. If the United States does not respond with incentives for U.S. industry to engage in developing more efficient, less polluting energy technologies, its competitiveness may diminish. The United States could wind up with an even less energy-efficient industrial base at home and a weaker trade position than it has today. A strong domestic approach would also enhance the political legitimacy of U.S.-led technological cooperation initiatives. As the world's largest consumer of energy, and one of its most inefficient, the United States has yet to demonstrate a commitment to reducing its own emissions of greenhouse gases.

THE COSTS OF PROMOTING TECHNOLOGICAL CHANGE

Our suggestions for institutional change, such as the cultivation of institutions that have a strong incentive to invest in energy-demand management, do not substitute for actual inputs of capital. Nor can improvements in indigenous technological capability, well-designed policy environments, or good products displace the need for investment in plant and equipment. The electricity sector provides an extreme example of the capital investment problems that face developing countries. According to World Bank estimates, electrical generating capacity will need to grow at over 6 percent per year in the 1990s, requiring capital investments of about $100 billion annually, in order to support a reasonable degree of economic growth. However, present trends suggest that only 10 to 12 percent of this funding is likely to be available from development banks and bilateral aid agencies. In the past, governments or private financing backed by government guarantees made up the balance. These sources diminished significantly during the 1980s and are unlikely to grow during the 1990s because of both the debt load of many developing countries and the highly inefficient and indebted state of many utilities.[37]

Wider use of energy-efficient technologies could alleviate this constraint somewhat, simultaneously reducing the rate of greenhouse gas emissions and capital requirements. For instance, although the construction of a plant for producing compact fluorescent lamps in India is estimated to cost about $10 million (largely in foreign exchange)—five to ten times the capital needed for a factory that would make a comparable number of incandescent bulbs—the substitution in lighting end use would yield capital savings of $3–5 billion in electric power capacity (about 40 percent of this in foreign exchange).[38]

These potential savings have a bearing on the hotly debated issue of additionality, that is, whether additional international transfers of funds, beyond those currently committed to development assistance, will be needed to stabilize or reduce the growth of greenhouse gas emissions from developing countries. Calculating additionality is tricky; one must first be clear about the desired level of emissions reductions and the baseline to which funds might be added. This baseline depends on assumptions about the path that would be taken by energy sectors in developing countries without interventions to control greenhouse gases. Moreover, a reasonable view on additionality requires consideration not only of the costs of hardware, but of the costs of developing technological capability and promoting technological change as well.

It is instructive to consider three different categories of energy technologies. In cases in which an energy service can be provided at a lower first cost and a lower life cycle cost while reducing greenhouse gas emissions, hardware spending should be below current projections. However, to put such technologies into use, additional aid may be necessary to provide technical assistance and institutional support. If the approach is seen as experimental and risky, financial incentives may be needed to get technology adopters to try something new.

Many other technologies have lower life cycle costs than those currently planned for use, but their first costs are higher. This situation is particularly true of many technologies used by consumers. The transaction costs of diffusing such technologies can be high, and consumers often must be subsidized to adopt them. In the near term, funds are likely to be needed to finance the higher first cost to the consumer. In addition, for this category of technologies, more technical assistance and institutional development funds are likely to be required than in the previous category because of the difficulty of developing institutions that can take a long-term perspective and that have an interest in diffusion.

Finally, there are technologies that command a premium for cleaner development. For instance, even with improvements in efficiency, energy systems in developing countries must continue to increase energy supplies to meet development goals. For many of these countries, the use of coal is much less expensive than are other fuels for electricity generation. Without spending some additional funds to cover the higher costs of a low-coal energy path, the international community will be hard-pressed to secure cooperation from such countries. Indeed, if lower demand for products that cause high–greenhouse gas emissions forces down prices, the premium may grow over time.

We must conclude, therefore, that technological cooperation for greenhouse mitigation will not be without costs for developed nations. Substantial additional funds will be needed to achieve significant reductions in greenhouse gas emissions. Environmentally unsound technology choices are not being made because the actors involved are irrational; most participants in the energy systems of developing countries have good reasons for their commitments to their current paths, and it could well be expensive to change their minds. However, if the momentum of system development can be shifted, a path with lower greenhouse emissions could become self-sustaining as improvements in technology, behavioral shifts, and institutional changes feed off each other. That momentum is the greatest promise that technological cooperation, understood in the broad sense that we have developed here, holds for the future.

ACKNOWLEDGMENTS

We thank Harvey Brooks, Henry Lee, and Brad Whitehead for their thoughtful comments. We also express our gratitude to the participants in efforts to promote the adoption of more environmentally sound technologies in developing countries for sharing their thoughts with us.

NOTES

1. Estimates of future emissions are scenario-dependent and are particularly sensitive to assumptions about growth and policies enacted to reduce greenhouse gas emissions. See Intergovernmental Panel on Climate Change, *Climate Change: The IPCC Response Strategies* (Geneva: World Meteorological Organization/UNEP, 1991); U.S. Environmental Protection Agency, Office of Policy Planning and Evaluation, *Policy Options for Stabilizing Global Climate* (Washington, DC: U.S. GPO, 1991); and U.S. Congress, Office of Technology Assessment, *Changing by Degrees: Steps to Reduce Greenhouse Gases*, OTA-O-482 (Washington, DC: U.S. GPO, February 1991). For a recent statement of the position offered in the text, see "Moves by Industrialized Nations Only Seen as Inadequate to Reduce Emissions," *International Environment Reporter* 16, no. 4 (24 March 1993): 205.
2. The climate convention negotiated at UNCED in Rio de Janeiro in 1992 makes explicit the responsibility of industrialized countries to "take all practicable steps to promote, facilitate and finance, as appropriate, the transfer of, or access to, environmentally sound technologies and know-how" to developing countries. It also states that industrialized countries "shall provide new and additional financial resources to meet the agreed full costs incurred by developing countries . . . in complying with their obligations" under the treaty. The specific commitments of the treaty, however, are for monitoring and information gathering; it does not specify particular goals for greenhouse gas emissions,

leaving ambiguous the degree to which industrialized countries will be expected to assist developing countries in actually achieving reductions. See report of the Intergovernmental Negotiating Committee for a Framework Convention on Climate Change on the work of the second part of its fifth session, held at New York, 30 April to 9 May 1992, A/AC.237/18 (Part II)/Add.1, May 15, 1992.

The implementation of the financial mechanism established by the climate convention was postponed until the first session of the Conference of Parties. Jousting about its design is ongoing; for example, see "GEF Restructuring Deemed Essential to Make It Financing Arm for Treaty," *International Environment Reporter* 16, no. 6 (24 March 1993): 198.

3. Mark D. Levine, Ashok Gadgil, Steven Meyers, Jayant Sathaye, Jack Stafurik, and Tom Wilbanks, "Energy Efficiency, Developing Nations, and Eastern Europe" (Washington, DC: Global Energy Efficiency Initiative/U.S. Working Group on Global Energy Efficiency, International Institute for Energy Conservation, June 1991), 22.

4. These lessons are similar to (though developed independently from) those put forward by Amulya K. N. Reddy in "Barriers to Improvements in Energy Efficiency," *Energy Policy* 19 (1991): 953–961.

5. For a review of development theory, see Amitav Rath, "Science, Technology, and Policy: A Perspective from the Centre," *World Development* 18 (1990): 1429–1443, especially 1432–1438. Of course, both modernization and dependency theories still have followers. For a dependency theory approach to the greenhouse problem, see Patrick McCully, "The Case against Climate Aid," *The Ecologist* 21 (November/December 1991): 244–249.

6. A. J. Gadgil and G. M. Jannuzzi, "Conservation Potential of Compact Fluorescent Lamps in India and Brazil," *Energy Policy* 19 (1991): 449–463; and Evan Mills, "Evaluation of European Lighting Programs," *Energy Policy* 19 (1991): 266–278.

7. For a listing and discussion of some of these programs, see the following two publications: Levine et al., *Energy Efficiency, Developing Nations,* 37–43; and U.S. Congress, Office of Technology Assessment, *Energy in Developing Countries* (Washington, DC: U.S. GPO, 1991), 38–39.

8. Only about 1 percent of the multilateral development banks' energy lending finances end-use energy efficiency. See Michael Philips, "The Least Cost Energy Path for Developing Countries: Energy Efficient Investments for the Multilateral Development Banks" (Washington, DC: International Institute for Energy Conservation, September 1991).

9. See, among many examples, Carl J. Dahlman, Bruce Ross-Larson, and Larry E. Westphal, "Managing Technological Development," *World Development* 15 (1987): 759–775; and Brent Herbert-Copley, "Technical Change in Latin American Manufacturing Firms: Review and Synthesis," *World Development* 18 (1990): 1457–1469.

10. Gatekeeping is not the exclusive domain of the Instituto de Investigaciones Eléctricas (IIE). Professionals within the national electric utility also participate in gatekeeping activities. Furthermore, the IIE is more than a gatekeeping institute. Building on its knowledge of state-of-the-art technology, it is involved in technological cooperation and technology innovation.

11. Richard R. Nelson, "What Is Public and What Is Private about Technology?" working paper 90-9, Consortium on Competitiveness and Cooperation, Center for Research in Management, University of California at Berkeley, 1990.

12. The importance of learning-by-doing is evident in case studies of Latin American firms. See *Technology Generation in Latin American Manufacturing Enterprises: Theory and Case-Studies Concerning Its Nature, Magnitude, and Consequences*, ed. Jorge M. Katz (London: Macmillan, 1987).

13. Martin Bell, "International Technology Transfer, Industrial Energy Efficiency, and Energy Policy in Industrializing Countries," in *CIFOPE/AIT/CEC Seminar on Energy Policy, Pattaya, Thailand, February 27–March 3, 1989*, ed. B. Lapillonne (Bangkok: Regional Energy Resources Information Center, Asian Institute of Technology, 1990).

14. Again, this point is borne out by case studies of Latin American firms. See Jorge M. Katz, *Technology Generation*. For more detailed discussions of the importance of an active technological strategy, see Dahlman et al., "Managing Technological Development"; and N. Chantromanklasri, "The Development of Technological and Managerial Capability in the Developing Countries," in *Technology Transfer in the Developing Countries*, ed. Mans Chatterji (New York: St. Martin's Press, 1990).

15. Dahlman et al., "Managing Technological Development," 760–762.

16. In the United States, for example, the EPA, the Agency for International Development, and the Department of Energy are constructing ENVIRONET, an information clearinghouse on environmental and energy-efficient technology. The UNEP is working on a similar system called the International Cleaner Production Clearinghouse.

17. National Academy of Engineering, *Cross-Border Technology Transfer to*

Eliminate Ozone-Depleting Substances (Washington, DC: National Academy Press, 1992), 26–29, 34–39, and 42–44. It is important to remember that the multilateral fund is relatively new, and many details of implementation remain to be worked out. For instance, developing countries have been reported to lack even the capability to apply for funds for some of the activities described in the text. In addition, wealthy nations have been slow to pay into the fund. See "Developing Countries Said to Face Severe Problems in Applying for Protocol Money," *International Environment Reporter* 15, no. 5 (11 March 1992): 127; and "New Target Dates Endorsed by Most Nations, Likely to Be Adopted in November, Tolba Says," *International Environmental Reporter* 15, no. 15 (29 July 1992): 492.

18. Dieter Ernst and David O'Connor, *Technology and Global Competition: The Challenge for the Newly Industrialising Economies* (Paris: OECD, 1989), especially 25–28 and 122–126.

19. We presume that enough biomass will be replanted in this project to ensure that the total carbon emissions are net zero.

20. "Energy from Biomass: A Growing Trend," *The Economist* 321, no. 7738 (21 December 1992): 107. For general information on this technology, see Joan M. Ogden, Robert H. Williams, and Mark E. Fulmer, "Cogeneration Applications of Biomass Gasifier/Gas Turbine Technologies in the Cane Sugar and Alcohol Industries" (paper presented at the Program for Acceleration of Commercial Energy Research Conference, New Delhi, 24–26 April 1990.

21. U.S. Congress, Office of Technology Assessment, *Fueling Development: Energy Technologies for Developing Countries*, OTA-E-516 (Washington, DC: U.S. GPO, April 1992), 67.

22. Madeleine Costanza, "Developing Countries and the Significance of Environmental Technology Transfer to an Agreement on Climate Change" (Paris: OECD Environment Directorate, Pollution Control Division, October 1990).

23. Francisco P. Bernardo and Gregorio U. Kilayko, "Promoting Rural Energy Technology: The Case of Gasifiers in the Philippines," *World Development* 18 (1990): 565–574.

24. The term "policy-assisted, market oriented measures" is drawn from Reddy, "Barriers to Improvements," 960.

25. Rene Bowser, "A History of the Montreal Protocol's Ozone Fund," *International Environment Reporter* 14, no. 23 (20 November 1991): 636–640.

26. The exact consequences of subsidies depend on elasticities that vary locally. For example, if low-priced electricity induces a shift away

from wood burning, which is highly inefficient, there may be a systemic improvement in energy efficiency due to electricity subsidies. We believe, though, that the more (and increasingly) common case is that low electricity prices discourage efficient use of energy.

27. The most recent account of BELLE is by Ashok J. Gadgil and M. Anjali Sastry, in "Stalled on the Road to the Market: Analysis of Field Experience with a Project to Promote Lighting Efficiency in India," in *Proceedings of the ACEEE 1992 Summer Study on Energy Efficiency in Buildings*, vol. 6 (Asilomar, CA, 30 August–5 September 1992), 6/57–6/70. (Also published by U.S. Department of Energy, report LBL-32447, Lawrence Berkeley Laboratory, Energy and Environment Division, October 1992). The authors comment that the experience, while it has not yet paid off in India, has led to programs in other countries, such as the recently announced ILUMEX program in Mexico, funded by the Global Environment Facility. Other reports on BELLE include Gadgil and Jannuzzi, "Conservation Potential"; and A. Gadgil, A. Rosenfeld, D. Arasteh, and E. Ward, "Advanced Lighting and Window Technologies for Reducing Electricity Consumption and Peak Demand: Overseas Manufacturing and Marketing Opportunities," report LBL-30389, U.S. Department of Energy, Lawrence Berkeley Laboratory, Energy and Environment Division, April 1991.

28. For instance, Howard S. Geller, "Establishing an International Energy Efficiency Agency: A Response to the Threat of Global Climate Change," *Energy Policy* 19, no. 7 (September 1991): 689–694; and "Global Change System for Analysis, Research, and Training (START): Report of a Meeting at Bellagio, December 3–7, 1990," report no. 15 of the International Geosphere-Biosphere Program of the International Council of Scientific Unions, Boulder, CO, 1991.

29. Jock R. Anderson, Robert W. Herdt, and Grant Scobie, *Science and Food* (Washington, DC: World Bank, 1988); and Douglas Horton, "Assessing the Impact of International Agricultural R&D Programs," *World Development* 14 (1986): 453–468.

30. David Jhirad, "Power Sector Innovation in Developing Countries: Implementing Multifaceted Solutions," *Annual Review of Energy* 15 (1990): 365–398.

31. While the organization of PACER brings commercial interests to the fore, it is not independent of political interests. The recent experience of the BELLE Project (see note 27 above) demonstrates that projects that seem economically and technically sound may be rejected due to institutional or political conflicts that can be expressed

within PACER. It was a PACER committee that blocked BELLE's access to foreign exchange. The lesson to be drawn is that such programs must walk a very fine line, empowering indigenous institutions so as to be effective, but working to develop mechanisms to prevent objections that are based entirely on bureaucratic considerations.

32. For a more detailed discussion of the role of brokers and bridging agents, see "Technological Cooperation for Sustainable Development," report of the Sixth Talloires Seminar on International Environmental Issues, Center for Environmental Management, Tufts University, 1991.

33. See note 2 above for discussion of the relevant provisions of the 1992 climate convention.

34. Initial efforts toward this goal are being pursued through an informal working group called the Multinational Working Group on Power Sector Innovations.

35. The Committee on Renewable Energy Commerce and Trade promotes U.S. renewable energy and energy efficiency technology by sponsoring trade shows, trade missions, and other activities that bring together U.S. federal agencies, industry, and potential overseas users.

36. The Overseas Private Investment Corporation is in the process of establishing an EIF to promote investments in environmentally sound technology overseas.

37. Gunter Schramm, "Electric Power in Developing Countries: Status, Problems, Prospects," *Annual Review of Energy* 15 (1990): 307–333. The World Bank recently put indebted utilities on notice. See "Bank Tightens Energy Loans," *Boston Globe*, 22 February 1993, 46.

38. Gadgil et al., "Advanced Lighting and Window Technologies," 5–6.

■ APPENDIX
Summary of the Report of IPCC Working Group I on Scientific Assessment

This appendix is a brief summary of the work of Working Group I of the Intergovernmental Panel on Climate Change (IPCC), a group of 170 scientists whose task was to assess available scientific information on climate change. Their work forms the scientific background to the policy strategies discussed in this book.

This summary of the IPCC's scientific assessment is necessarily much less detailed than the original report; we recommend that interested readers consult the actual reports for more information (see J. T. Houghton, G. J. Jenkins, and J. J. Ephraums, eds., *Climate Change: The IPCC Scientific Assessment* [New York: Press Syndicate of the University of Cambridge, 1990]).

HUMAN ACTIVITIES AND CLIMATE CHANGE

The earth's climate is a function of the amount of solar radiation it absorbs, reflects, and emits. The planet intercepts short-wave radiation from the sun; some of this energy is immediately reflected by the earth and the atmosphere, and some of it is absorbed by the atmosphere, plants and animals, oceans, land, and ice. The planet reemits energy in the form of long-wave, invisible, infrared radiation. Some long-wave radiation leaving the earth is absorbed by the blanket of greenhouse gases in the troposphere and stratosphere and reemitted into the cooler atmosphere above. These gases include water vapor, carbon dioxide, methane, nitrous oxide, ozone, and CFCs. This process is what is called the greenhouse effect, and without it the planet would be cooler by about 33° Celsius.

Human activities, especially over the past hundred years, have increased the atmospheric concentration of greenhouse gases. Carbon dioxide emissions are the greatest contributor to the enhanced greenhouse effect. Since the early nineteenth century, the combustion of fossil fuels and accelerated deforestation has led to an increase of 26 percent in atmospheric carbon dioxide. Emissions of CFCs did not begin until the

1930s, when they were invented. Methane concentrations have more than doubled because of agricultural practices, coal mining, and venting of natural gas. Nitrous oxide has increased by about 8 percent since preindustrial times; scientists are uncertain of the major sources. Atmospheric concentrations of long-lived gases, such as carbon dioxide, CFCs, and nitrous oxide, adjust only slowly with changes in emissions. The longer emissions continue to increase at present-day rates, the greater the reductions will need to be to stabilize atmospheric concentrations at a given level.

Global mean surface temperature has increased by 0.3° to 0.6° Celsius over the past hundred years, and the sea level has increased by 10 to 20 centimeters over the same period. While this amount of warming is broadly in keeping with the predictions of climate models, it is also possible that these increases could be due to natural climate variability. An unequivocal finding of an enhanced greenhouse effect caused by human activities is thought to be at least a decade away.

FEEDBACKS TO WARMING

The earth and its atmosphere form a very complex system, and there are feedbacks to the greenhouse effect that could significantly modify future concentrations in a warmer world. The net emissions of carbon dioxide from terrestrial ecosystems could increase if large plant systems cannot adjust to climate changes or if higher temperatures increase plant respiration at a faster rate than photosynthesis. On the other hand, a net negative feedback effect is possible if a higher concentration of atmospheric carbon dioxide results in enhanced productivity of natural ecosystems. The extent to which ecosystems can sequester increasing atmospheric carbon dioxide is still unknown. Warmer oceans may take up less carbon dioxide because of changes in the chemistry of carbon dioxide in seawater, in biological activity, or in the rate of exchange of carbon dioxide between the surface layers and the deeper waters of the ocean. These feedback processes are still poorly understood.

THE IPCC SCENARIOS

Four scenarios of future greenhouse gas emissions were developed by the IPCC. The business-as-usual scenario (Scenario A) assumes that few or no steps will be taken to curb human-made greenhouse gas emissions. In Scenario B, the energy supplies are shifted to low-carbon fuels, large efficiency increases are realized, deforestation is reversed, and the Montreal Protocol is implemented. In Scenario C, a shift toward renewable energies and nuclear energy takes place in the second half of the next century. CFCs are phased out, and agricultural emissions are limited. Scenario D assumes a shift to renewable energies and nuclear energy in the first half of the century. This scenario, which assumes stringent controls in industrialized countries combined with moderated emissions growth in developing countries, could stabilize atmospheric concentrations. In Scenario C, carbon dioxide emissions are reduced to 50 percent of 1985 levels by the middle of the next century.

RESPONSE OF EARTH SYSTEMS TO THESE CHANGES

If the concentration of atmospheric greenhouse gases increases, the atmosphere's capacity to capture and hold heat will increase. This phenomenon should result in an increase in the earth's average temperature, which in turn could alter precipitation levels in various regions and during different seasons. It should also result in an increase in the rate of sea rise. Models predict that surface air will warm faster over land than over oceans, and minimal warming will occur around the polar regions. In high northern latitudes in winter, warming is predicted to be 50 to 100 percent greater than the global mean, and in regions of sea ice in the summer, warming is predicted to be substantially smaller than the global mean. Precipitation is predicted to increase, on average, in middle- and high-latitude continents in the winter.

Under the business-as-usual scenario, the IPCC predicts a likely increase in global mean temperature of about 0.3° Celsius per decade, resulting in an increase of about 1° Celsius above the present value by 2025 and 3° Celsius before 2100. Global average precipitation and evaporation would increase by a few percentage points by 2030. Areas of snow and ice are expected to diminish. Global mean sea level is expected to rise about 6 centimeters per decade, or about 20 centimeters by 2030 and 65 centimeters before 2100, because of the thermal expansion of the oceans and the melting of some land ice.

Under emissions scenarios that assume progressively increasing emissions controls, global mean temperature is predicted to rise about 0.2° Celsius in Scenario B, just above 0.1° Celsius in Scenario C, or about 0.1° Celsius in Scenario D per decade. By 2100 sea level is expected to rise by just under 50 centimeters in Scenario B, about 40 centimeters in Scenario C, or above 30 centimeters in Scenario D. Even if we were immediately able to stabilize greenhouse gas emissions at present-day levels, the temperature would rise by about 0.2° Celsius per decade and sea level would also continue to rise for the next few decades.

These predictions are for climate change on a global scale, which is not useful for determining its actual impacts. Temperature and sea level will not rise uniformly over the globe; there will be significant regional variations. It was the judgment of the IPCC working group that global models cannot yet yield reliable regional predictions at the level needed for impact assessments, especially for changes in precipitation and soil moisture. Climate models offer no clear prediction that weather variability will change in the future or that storms will increase in a warmer world.

CLIMATE CHANGE MODELS AND WHAT CAN BE LEARNED FROM THEM

Different greenhouse gases alter the radiative balance (the amount of energy that is absorbed, emitted, and reflected) of the planet to differing degrees, depending on many factors, including the absorption strength of the particular gas molecule and its concentration and expected lifetime in the atmosphere. For purposes of climate modeling, these factors are usually combined in an expression of global warming potential relative to carbon dioxide.

Currently, the most highly developed set of tools for modeling global climate change is the general circulation models, which were derived from weather forecast models. To make a climate forecast, the climate model is run for a few (simulated) decades, and the output is compared with the climate of the real atmosphere. This exercise is then repeated with different concentrations of greenhouse gases. These models have generally been run coupled with models of the upper ocean to investigate climate change. The IPCC constructed its various scenarios of future climate change using simplified versions of these tools. Only a few models linking all the components of the climate system have been developed to date, largely because of the enormous demands on computer time required.

Significant modeling uncertainties come into any climate model predictions because scientists do not fully understand the feedback mechanisms, the responses of sources and sinks of various gases, and the action of oceans, clouds, and polar ice caps.

■ INDEX

Acid rain, 185–86
Adaptability of policy instruments, 199–200
Adaptation investments, 2
Adaptation of technology, 271–72
Additionality issue, 281–82
Adler, Emmanuel, 58
Administrative flexibility/adaptability, 230–31
Ad valorem tax on fuel use, 248–49
Advisory groups, 69
Ambient permits, 182–83
Analyzing compliance information, 125–26
Antarctic ozone hole, 52
Appliance standards, 266
Asymmetries of interest, 85–87, 90–93
Australia, 44

Banks, development, 266–67, 279
Base case simulation, 256–57
Basel Convention on Transboundary Movement of Hazardous Wastes, 83
Baseline emissions growth, 251
Baseline protocol, 61
Best available control technology (BACT), 221
Bilateral agreements, 20
Blocking coalitions:
CFC negotiations, 51–52
disparate interests unified into, 59–60
EU, 11
extent of, 52–56
ideological tenor, 69–70
incremental approaches, 60–63
Law of the Sea treaty, 48–51
linkages among issues/protocols, 63–67
North-South polemics, 67–68
small-scale expanding agreements, 70–72

traditional control regimes, alternatives to, 57–59
Bombay Efficiency Lighting Large-Scale Experiment (BELLE), 275
Boskin, Michael, 42
Brazil, 12. 45, 53, 269
Brooks, Harvey, 7
Brundtland Commission, 70
Btu (British thermal unit), 233, 239
Bush, George, 26, 42, 150
Business Council for Sustainable Development, 69

Canada, 44, 53, 153
Capital formation/investment in general equilibrium model, 255
Carbon dioxide, 1, 5, 289
costs for reducing, 53
cross-national comparison estimates, 90–93
offsets, 22–23, 31–32
prices, low energy, 10
targets and timetables for reducing emissions, 81
Carbon taxes, 219, 237
administrative flexibility/adaptability, 230–31
alternatives to, 248–49
base case simulation, 256–57
Clean Air Act of 1970, 220–23
conclusions on, 251–52
developed countries, cooperation among, 226–27
distributional effects, 249–50
domestic *vs.* international, 130
economic cost of stabilizing climate change, 238–45
environment *vs.* cost, 225
EU, 55
general equilibrium model, 252–56
IEF, hypothetical, 224–25